EXPLORATIONS IN PHYSICS

AN ACTIVITY-BASED APPROACH TO UNDERSTANDING THE WORLD

EXPLORATIONS IN PHYSICS

AN ACTIVITY-BASED APPROACH TO UNDERSTANDING THE WORLD

DAVID P. JACKSON
Dickinson College

PRISCILLA W. LAWS
Dickinson College

SCOTT V. FRANKLIN
Rochester Institute of Technology

With Contributing Author:
KERRY P. BROWNE
Dickinson College

John Wiley & Sons, Inc.

ACQUISITIONS EDITOR	Stuart Johnson
SENIOR PRODUCTION EDITOR	Elizabeth Swain
SENIOR MARKETING MANAGER	Bob Smith
COVER DESIGNER	Karin Gerdes Kincheloe
ILLUSTRATION EDITOR	Anna Melhorn
PHOTO EDITOR	Hilary Newman and Elyse Rieder
COVER PHOTO	©Eyewire

This book was typeset by the authors and was printed and bound by Courier Westford. The cover was printed by Phoenix.

This book is printed on acid-free paper.

To order books or for customer service, call 1(800)-CALL-WILEY (225-5945).

ISBN 0-471-32424-8

Printed in the United States of America

10 9 8 7 6 5 4 3 2 1

PREFACE

Explorations in Physics (EiP) is a set of curricular materials developed to help non-science majors acquire an appreciation of science, understand the process of scientific investigation, and master concepts in selected topic areas. Although *EiP* contains both text and experiments, it is neither a textbook nor a traditional laboratory manual. Rather, it is an *Activity Guide*—a student workbook that combines text with guided inquiry activities. Each *EiP* unit covers a specific topic area and has been designed as a stand-alone unit to give instructors maximum flexibility in designing courses. Whenever possible, each student concludes the investigations in a unit by undertaking an extended scientific project designed and carried out with two or three fellow students.

Explorations in Physics represents a philosophical and pedagogical departure from traditional modes of instruction. While courses for non-science majors have typically focused on intrinsically interesting topics to help motivate students, the primary format of these classes remains the lecture. *Explorations in Physics* attempts to enrich the study of interesting topics with a hands-on approach to learning. Careful attention has been paid to maintaining a focused "story line" that directs students to connect specific activities to real-world phenomena. This provides a framework for students to develop their own scientific investigations which they complete during the project phase of the course.

A major objective of *EiP* is to help students understand the basis of knowledge in physics as a subtle interplay between experiment and theory. Students spend most of their class time making predictions and observations in order to develop coherent conceptual models of various phenomena. There are two major reasons for emphasizing the process of scientific investigation. First, introductory science students frequently lack the conceptual understanding of physical phenomena necessary to comprehend theories and mathematical derivations presented in lectures. Second, students who are actively involved in scientific investigations show a high level of commitment to the activities and a greater appreciation of the topics they are studying.

In designing *EiP*, we have taken advantage of the results of physics education research on student learning and attitudes. We have used findings from other researchers as well as our own surveys and student interviews to shape the curricular materials. Our assessments of student learning indicate that students who complete a sequence of *EiP* activities in supportive learning environments achieve significant improvements in their conceptual understanding of the topics studied.

TOPICS COVERED AND THE MODULAR FORMAT OF THE UNITS

Explorations in Physics organizes topics into individual units. The core material in these units is explored through a series of activities that students work through in peer learning groups. These activities consist of predictions, observations, measurements, analysis, and reflections, and are designed to guide the students through the process of scientific inquiry. The core material for a unit takes about 18 class hours to complete and is typically followed by a student-directed project.

These units are designed to be completely independent of one another and can be introduced in any order. This flexibility allows instructors to design a course to match their particular needs and interests. Four *Explorations in Physics* units are contained in this volume. These include:

- *Unit A: Force, Motion, and Scientific Theories*
- *Unit B: Light, Sight, and Rainbows*
- *Unit C: Heat, Temperature, and Cloud Formation*
- *Unit D: Buoyancy, Pressure, and Flight*

Additional units are in various stages of development. They include *Unit E: Energy, Fuels, and the Environment, Unit F: Patterns, Fractals, and Complexity,* and *Unit G: Sound, Vibrations, and Musical Tones.* These units are available upon request to the authors.

USING THE ACTIVITY GUIDE IN VARIOUS INSTRUCTIONAL SETTINGS

Explorations in Physics was originally designed to be used with relatively small classes in a Workshop/Studio setting that combines laboratory and computer activities with discussions. The materials were tested and refined over a 7-year period at Dickinson College, Santa Clara University, and Rochester Institute of Technology. The schedule for *EiP* courses was different at each of these institutions, and the number of topics covered and the balance between guided inquiry and projects had to be adjusted accordingly. Some common implementations are described in Table 1. In most cases, the suggested schedules also allow extra days for exams, review sessions, and oral project presentations.

Academic Calendar	Class Schedule	Core Material	Student Projects
Semester	3 hrs/week	1 Unit	1 Full Project
Semester	6 hrs/week	2 Units	2 Full Projects
Quarter	3 hrs/week	1 Unit	1 Shortened Project
Quarter	6 hrs/week	2 Units	1 Full Project

Table 1: Common implementation schedules for core materials and projects

We recognize that not all institutions have the resources to provide a Workshop learning environment in which lectures and labs are combined. As outlined below, these materials can also be adapted for use in more traditionally structured classes.

Traditional Lecture Sessions: It is possible to incorporate individual activities into lectures as demonstrations, similar to Interactive Lecture Demonstrations developed by David Sokoloff and Ronald Thornton. In these demonstrations, students record their predictions, discuss them with fellow students, and then watch as the instructor performs an experiment. Questions in the activity guide lead students to reconcile their predictions and observations.

Traditional Lecture Sessions with Laboratory: In cases where a complete unit is introduced into a traditionally scheduled course, the labs and lectures can be coordinated so students can work through the unit in sequence with some activities being done as interactive lecture demonstrations and others as laboratory exercises.

COMPUTER TOOLS AND STUDENT PROJECTS

When used properly, computers can greatly enhance student learning. In *Explorations in Physics*, we use computer-based laboratory tools for the real-time collection and graphing of data. Data is collected through sensors that are connected to a computer via an interface. Available sensors are capable of measuring a variety of physical quantities such as force, motion, temperature, light intensity, and pressure. These sensors, interfaces, and software are available from many vendors including PASCO Scientific and Vernier Software.

Student-directed projects are one of the most exciting elements of *EiP*. These projects are carried out both during and outside class time and culminate in group oral presentations and individual written reports. These projects enable each group of students to investigate a topic of their own choosing. This helps to reinforce their understanding of the core material while giving them first-hand experience with the process of scientific investigation. For these reasons, we recommend dedicating equal amounts of class time to projects and core materials. While individual circumstances may preclude this, we cannot overemphasize the value of projects. At the end of each unit we have included descriptions of some viable student projects. It should be stressed, however, that these are only suggestions. The most successful projects are often those that students develop for themselves based on personal interests.

STAYING UP TO DATE

The authors have been offering workshops on various aspects of teaching *Explorations in Physics* including weeklong summer workshop and shorter workshops offered at national AAPT meetings. A schedule of upcoming workshops will be posted on the *EiP* website. The *Explorations in Physics* website also contains instructor materials for each unit. These include tips to help instructors with activities and equipment, sample syllabi, homework assignments, examinations, and other course documents. The web address is: http://physics.dickinson.edu/EiP. If you have problems logging on, call the *Workshop Physics* Project Office at (717) 245-1845 between 8:00am and 5:00pm EST.

ACKNOWLEDGMENTS

This project began in the fall of 1994 with a grant from the Charles A. Dana Foundation. Since 1994 the project has benefited from the expert advice and support of numerous individuals. We are especially grateful to Uri Treisman, Senior Advisor to the Dana Foundation's Education Program, for his creative work in helping us launch this project. Feedback from an Advisory Committee that met during the early stages of this project was of paramount importance. Members of this committee included: Nancy Baxter-Hastings, Robert Boyle, Michael Burns-Kaurin, Michael Chabin, Nancy Devino, Robert Fuller, Gerald Hart, Sandra Melchert, Hans Pfister, Dick Stanley, Carol Stearns, Ronald Thornton, and Sheila Tobias.

In addition, a substantial number of ideas have been derived from casual conversations and published articles. It is not possible to list everyone who has influenced the development of these materials, but some of the more important contributions came from Lillian McDermott and the Physics Education Group at the University of Washington, E.F. "Joe" Redish and the Physics Education Research Group at the University of Maryland, David Sokoloff at the University of Oregon, and Ronald Thornton at Tufts University.

We have received valuable comments from a number of reviewers and Beta Testers. These include Robert Beichner (North Carolina State University), John Carini (Indiana University), John Christopher (University of Kentucky), Jeff Collier (Bismarck State College, ND), Lynn Cominsky (Sonoma State University), Bob Fuller (University of Nebraska), Donald Greenberg (University of Alaska, Southeast), Bob Bogar and Pat Keefe (Clatsop Community College), Gerald Hart (Moorhead State University), Harold Hart (Western Illinois University), Charles Hawkins (Northern Kentucky University), Mark Lattery (University of Wisconsin, Oshkosh), Marie Plumb (Jamestown Community College), Patricia Rankin (University of Colorado), David Lee Smith (LaSalle University), Chuck Stone (Forsyth Technical Community College), Beth Ann Thacker (Texas Tech University), Bob Tyndall (Forsyth Technical Community College), Mark Winslow (Independence Community College), and Gail Wyant (Cecil College).

We have also benefited from the many ideas contributed by our colleagues in the physics and astronomy departments at Dickinson College, Santa Clara University, and Rochester Institute of Technology. In addition, Kerry Browne has made significant administrative, intellectual, and artistic contributions to this project. In designing illustrations and refining storylines he has been instrumental in bringing these materials to their final published form. We are also grateful for the administrative support of Gail Oliver, Maurinda Wingard, and Sara Buchan. We would be remiss if we did not mention the many generations of students whose continual feedback over the years has helped make these materials more student friendly. Specifically, we would like to thank those students whose efforts went above and beyond our expectations and whose genuine interest and creativity have helped us learn new things about physics.

This project would never have been published were it not for the hard work and enthusiasm of our Acquisitions Editor, Stuart Johnson, and the staff at John Wiley and Sons, Inc. We gratefully acknowledge the funding agencies whose support has led to the publication of these materials. These include the Charles A. Dana Foundation, the Fund for the Improvement of Post Secondary Education (FIPSE), and the National Science Foundation, including the NSF PFSMETE program that enabled Scott Franklin to join the project. We have also received generous financial support from Dickinson College, Santa Clara University, and Rochester Institute of Technology. Lastly, we would also like to thank our families for their continued patience and support throughout the life of this project.

David P. Jackson, *Dickinson College*
Priscilla W. Laws, *Dickinson College*
Scott V. Franklin, *Rochester Institute of Technology*
April 2002

ABOUT THE AUTHORS

DAVID JACKSON

David Jackson received his bachelor's degree from the University of Washington in 1989 and his Ph.D. from Princeton University. After completing his Ph.D. in magnetic fluid pattern formation in 1994, he joined the faculty at Dickinson College as an assistant professor and became the project director for *Workshop Science*. In this latter capacity, he collaborated with Priscilla Laws on design of a year-long physical science curriculum for non-science majors and pre-service teachers which forms the basis of the materials contained in this volume. In 1997, he joined the physics faculty at Santa Clara University as an assistant professor. There, he served as a principal investigator for the *Workshop Science* project and developed a research program to explore how multiple domains affect the pattern formation process. As part of this responsibility, he has developed and adapted the *Workshop Science* curriculum for use at Santa Clara. In 2001, he returned to Dickinson College where he continues to develop curriculum and investigate pattern formation processes courses as an assistant professor. Professor Jackson is an active member of the American Physical Society and the American Association of Physics Teachers.

PRISCILLA LAWS

Priscilla Laws received her bachelor's degree from Reed College and a Ph.D. from Bryn Mawr College in theoretical nuclear physics. She has been a faculty member at Dickinson College for many years. She is the author of articles and books on the health effects of medical and dental x-rays, the impact of energy use on the environment, and activity-based physics teaching. As part of the *Workshop Physics* project which she initiated in 1986, she has developed curricular materials, apparatus and computer software and hardware used in introductory physics teaching.

Dr. Laws has received awards for software design and curriculum innovation from EDUCOM/NCRIPTAL, Computers in Physics, the Sears-Roebuck Foundation, and the Merck Foundation. In 1993, she received the Dana Foundation Award for Pioneering Achievement in Education with Ronald K. Thornton and in 1996, the American Association of Physics Teachers bestowed the 1996 Robert A. Millikan Medal to Professor Laws for notable and creative contributions to the teaching of physics. She has been a principal investigator on a number of curriculum development projects funded by FIPSE and NSF. In 1994 she received a seed grant from the Dana Foundation to begin development of these *Explorations in Physics* units as part of the *Workshop Science* project.

SCOTT FRANKLIN

Scott Franklin received his bachelor's degree from the University of Chicago in 1991 and his Ph.D. from the University of Texas at Austin in 1997. He is currently on the faculty at Rochester Institute of Technology, where he is adapting *EiP* for use at a technological institute. Scott joined the *Workshop Science* project in 1998 with an National Science Foundation Postdoctoral Fellowship in Science, Mathematics, Engineering, and Technology Education (SMETE). Under the mentorship of Priscilla Laws, Scott developed new units, extensively revised existing units, and performed basic educational research on student conceptions and attitudes which were crucial to *EiP*'s successful development and implementation. Scott remains active in the Physics Education Research community, where he is currently co-editor of the Proceedings of the AAPT Physics Education Research Conference. In addition to curriculum development and physics education research, Scott maintains a lab investigating topics in nonlinear dynamics, including granular materials and dislocation dynamics.

TABLE OF CONTENTS

UNIT A

FORCE, MOTION, AND SCIENTIFIC THEORIES

DETAILED CONTENTS

UNIT A

FORCE, MOTION, AND SCIENTIFIC THEORIES

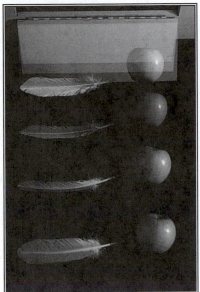

©James Sugar/Black Star

"Of all the intellectual hurdles which the human mind has confronted and has overcome in the last fifteen hundred years, the one which seems to me to have been the most amazing in character and the most stupendous in the scope of its consequences is the one relating to the concept of motion."

–Herbert Butterfield

0   OBJECTIVES

1. To understand how a scientific theory relating forces and motion can be developed from systematic observations and experiments.

2. To observe, classify, and describe one-dimensional motion using different representations including words, pictures, graphs, and mathematical equations.

3. To observe how forces affect the motion of an object that is moving along a line.

4. To use your observations to develop specific hypotheses about motion and forces and to design experiments to test these hypotheses.

5. To apply your ideas about forces and motion to the phenomena of gravity and friction.

6. To learn more about the nature and causes of motion and the process of scientific research by undertaking an independent investigation.

0.1   OVERVIEW

Why Study Motion?

We are surrounded by a world in motion. It is difficult to think of a single phenomenon that does not involve it. Motion is so ubiquitous, and your experiences with it so numerous, that it is impossible for you not to have some ideas about its causes. Consider the following questions. If you push on a box with a constant force, how will it move? How will the box move if you were to push twice as hard? Indeed, what exactly does it mean to push "twice as hard?" Even if you are not able to give quantitative answers to these questions, you probably have an intuitive feeling for how the box would move. This feeling comes from years of observations and hands-on experience with pushing and pulling on objects. These experiences helped you form an understanding of motion and its causes that enable you to walk, run, jump, and drive a car. Since these are all things you can do without thinking, you might even be wondering what you can possibly learn about motion that you don't already know.

The major reason why we are asking you to undertake a careful study of motion is because you are already very familiar with it. The study of motion will provide you with a straightforward example of how scientific ideas develop. Your ideas regarding motion are *hypotheses*—deductions based upon informal observations. You may be surprised to find that some of your hypotheses, which are based on your previous experience with motion, fail to provide a unified and simple scheme that can describe some common situations. For example, imagine a heavy safe that must be pushed across a carpeted floor. One person pushes on the safe, but it does not budge. A second person is enlisted to help and the safe moves across the floor at a slow, steady speed. Add a third person, however, and the safe now races across the floor quite easily. In fact, with three people pushing, it may be difficult to control the motion of the safe as it speeds up. So even in this very simple situation, pushing on a safe with one, two, or three people can lead to different kinds of motion (no motion, slow steady motion, fast and somewhat uncontrolled motion). Can your ideas of motion provide a clear explanation for why each of these safe motions is different?

Figure A-1: One person pushing a safe, two people pushing a safe, three people pushing a safe.

It is tempting to develop three different explanations to describe the safe motions. First, small forces result in no motion at all; second, intermediate forces result in slow, steady motion; and third, large forces result in more rapid and uncontrolled motion. This set of explanations seems to describe these three situations just fine. Is there anything wrong with using three explanations to describe these three situations? Well, not exactly. But as we observe additional situations involving motion, we are forced to develop more and more separate explanations. Ultimately, our view of the world would consist of a huge number of disjointed ideas. A single idea that explains all motions, would be more useful.

In cases such as these, scientists often use a principle known as *"Occam's Razor,"* named after the English Philosopher, William of Occam (1285-1347/49). Occam's Razor states that if you have two competing ideas, each of which agrees with observation, then the simpler one is preferable. After all, why would you want to use a more complicated description of something when a simpler one works just as well? Sometimes a complicated description is necessary, but only if no simpler one can explain all the observations. How, exactly, do we search for simple descriptions? By making systematic observations of different types of motion and modifying our hypothesis so that it is consistent with all of them.

Activity 0.1.1 Hypothesizing About Safe Motions

a) Recall the description of the safe that doesn't budge with a single pusher, moves along a line at a constant speed with two pushers, and moves more rapidly and less predictably under the influence of three pushers. Can you think of a single hypothesis that would enable you to explain all three of these motions?

If you had trouble answering the question just posed, that's all right. A major goal of this unit is to enable you to develop a theory of motion based on experimental observations that will allow you to construct a satisfactory answer to this question.

Scientific Theories and the Laws of Motion

The study of motion dates back thousands of years, but it wasn't until the 17th century that Sir Isaac Newton (1642-1727) proposed a satisfactory description of motion. Why did it take so long to understand something as apparently simple as motion? That is a difficult question. A major impediment to developing a good understanding of motion was the lack of well-defined experimental methods for making observations. Galileo Galilei (1564-1642) is often called the father of experimental science as a result of his use of careful experiments that enabled him to gain insights into the nature of motion. The concept of evidential support is a theme that will recur throughout this course. Scientific *theories* are based on evidence, and what makes a theory *scientific* is the fact that it is open to falsification. This means that scientific theories make predictions that can be tested experimentally. Thus, a scientific theory can never be *proven* correct, but it is always possible that a future observation will disagree with the theory. When this happens, a new theory is sought that can explain all the observations. In this way, scientific ideas are in a continual state of improvement and verification.

A hypothesis that can explain a large body of observational evidence is called a theory or sometimes a law. Traditionally, scientific laws are supported by more evidence than are scientific theories. Nevertheless, sometimes a scientific law is replaced by a more

Hypothesis $\longrightarrow$ Many Tests and Refinements $\longrightarrow$ Theory or Law
(Untested explanation)   (Partially tested explanations)   (Well tested explanation)

accurate theory. For example, Einstein's Theory of Relativity has superseded Newton's

Universal Law of Gravitation because it can explain a larger number of experimental observations. This should be kept in mind when you encounter these terms.

Activity 0.1.2 Hypotheses and Theories

a) Explain in your own words what the difference between a hypothesis and a theory is.

The Activities in this Unit

In this unit, you will learn about the scientific process as well as about motion. You will make assumptions, ask questions, build hypotheses, and devise experiments to test your hypotheses. The first step in understanding motion is to learn to describe it using multiple representations including words, graphs, and equations. You will also need to make measurements and to understand what these measurements mean. Contemporary computer tools facilitate the process of recording and displaying data. These will be used throughout the course as you study different phenomena.

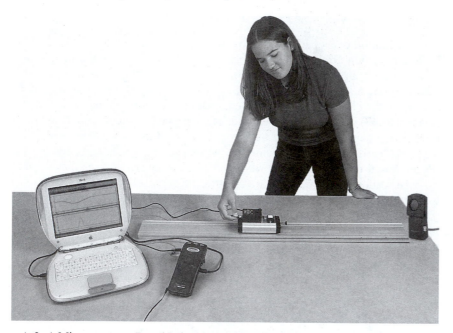

Figure A-2: A Microcomputer Based Laboratory (MBL) set-up, including force and motion sensors. (Courtesy Vernier Software & Technology)

You will begin by trying to describe the motion of someone riding on a low friction cart under the influence of a steady push. This will lead to a more careful consideration of how to describe simple one-dimensional motions. Next you will learn about how forces are defined and measured. Then, you will be ready to revisit the question of how forces acting along a line influence the speed of a moving object. This will allow you to develop a tentative hypothesis about the relationship between the forces acting on an object and the motion of that object. You will continue to refine your understanding of the concept of force and make a number of additional observations relating force to motion. Your activities should enable you to develop a comprehensive theory of motion that provides a single unified explanation for the many motions you will observe.

Note: The classical laws of motion that we will develop in this unit provide, for all practical purposes, "exact" descriptions of the motions of everyday objects traveling at ordinary speeds. During the early part of the twentieth century two new theories were developed that turned out to provide more complete descriptions than the classical laws of motion—*quantum mechanics*, which describes the motion of very small objects, and *relativity*, which describes objects that move extremely fast. The investigation of these descriptions is beyond the scope of these materials.

One final note. Since the theory of motion you develop will be based on observations of simple systems, you might wonder whether these theories should apply to more complicated motions. Scientists often work with the assumption that, until proven otherwise, simple theories *do* apply to more complicated situations. This assumption must always be tested, and such tests lead to predictions about new experiments.

1	**THE MEASUREMENT PROCESS**

A good way to begin a careful study of motion, is to try to describe simple motions. But what makes a motion simple? There are many different kinds of motions, and if we are going to choose the simplest, we must first figure out how to *describe* them. Describing motion helps us look for similarities and differences, and allows us to categorize them according to their complexity. From there, we can begin our study of motion by looking at the simplest cases.

You will need some of the following equipment for the activities in this section:

- Kinesthetic cart or other large, wheeled "vehicle" [1.1]
- Heavy ball (such as a bowling ball) [1.1]
- Meter stick [1.1]
- Stopwatch [1.1]
- Ruler [1.1]
- Beanbags [1.1]
- Masking tape [1.1]
- Graph paper [1.1]

1.1   A FIRST LOOK AT MOTION

Since we want to learn about how forces affect motion, why not just push on an object and observe how it moves. That seems reasonable enough. Let's give it a try! In the following activity you will push on an object that can roll freely and attempt to measure its motion.

Figure A-3: A person riding on a low-friction cart being pushed.

Activity 1.1.1 The Motion of a Pushed Object

a)   Without consulting your partners, take a stab at describing the *motion* of an object.

b) Now consult with your partners and devise a method for measuring the motion of a pushed object. Explain your procedure.

c) Again, consult with your partners and explain how you will maintain a constant *push* on the object.

d) Now it's time to perform the experiment. One of you should sit on a skateboard or low friction cart while someone else gives the rider a reasonably constant push. Try to measure the rider's motion and summarize your results below.

e) Was it actually possible to measure the *motion* of the rider? Explain briefly.

Although this seems like a simple experiment, you probably found that it was not that easy to carry out. For example, to maintain a constant push, you probably had to rely on your sense of touch in some way. This means the results are subjective rather than objective. Furthermore, it is not even obvious exactly what measuring *motion* means. One group might decide to measure the speed of the object while another might try to measure the object's position. Again, this is somewhat subjective. In order to come to a more solid understanding of motion, it is important to develop some common definitions so that we can all agree on what it is we are actually trying to measure.

Investigating Different Types of Motion

Now, before we can build these definitions, we must first make sure that all of us have experienced the same motions. Your instructor will demonstrate a number of different motions. These may include tossing an object in the air, rolling a ball on a level floor, sliding a flat object across a level surface, rolling a cart on an flat surface and on an inclined surface. Your job will be to describe the motions and then to classify them. Keep that in mind as you are observe each motion.

Activity 1.1.2 Describing and Classifying Motions

a) Briefly describe each of the demonstrated motions. (Just try to give the main features without being too detailed.)

Describe

b) Now, classify these motions into two or three (or more, if needed) different categories. Are there other ways of classifying them?

Classify

c) Which of the above categories of motion appears to be the simplest? Which do you consider the most complicated?

Simplify

Describing Constant-Speed Motion

In order to build an understanding of motion, let's begin by looking at the simplest type of motion we can think of. This is something that scientists often do when they are faced with a phenomenon. First, they try and understand a very simple situation, and then they move on to more complicated situations.

Most people agree that one of the simplest possible motions is when an object moves in a straight line without speeding up or slowing down. Some people might call this steady motion while others might say it is a uniformly moving object. We will use the term constant-speed motion to describe this situation. You saw some examples of this kind of motion in the last activity. We will now attempt to expand our understanding of this simple type of motion by trying to describe precisely what it means to move with a constant speed. This next activity introduces us to this concept.

Activity 1.1.3 Defining Speed

a) With your partners, try to define the term *speed* without using the word speed (or anything relating to speed, such as velocity, speeding up, slowing down, etc.) in your definition. Finish the sentence, "An object is moving with a constant speed if…"

If you found that you weren't quite sure how to answer the questions posed above, don't worry. Most people can correctly identify an object that moves with a constant speed, but many are unable to explain precisely how they know it's moving with a constant speed except to say that its speed doesn't change. Since we are ultimately interested in studying what causes more complicated motions, it is essential that we learn to describe this simple motion first. Only then will you have a basis for understanding more complex motions.

As you consider more complicated motions, you will see that it becomes increasingly difficult to describe them with words. Thus, you will be introduced to different ways of describing the same motions, some of which are very compact. For example, graphical displays of data can carry a lot of information that may be difficult to describe with words. Because of this, it is important that you learn to read and understand the information displayed in a graph.

The following activity asks you to do one thing: measure the speed of an object that moves with a constant speed and explain how you know that its speed is not changing. A heavy ball rolled on a hard, level surface has a nice constant-speed motion. Your instructor may have some specific instructions on how and where to perform this experiment.

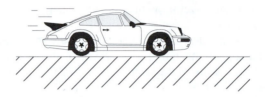

Figure A-4: In a car, the speedometer indicates how fast the car is moving. How can we measure speed in the lab?

Activity 1.1.4 What is Speed?

a) Find a place where you can roll your ball (not too slowly) on a hard, level surface. Then, use any of the materials available to you to measure the object's speed. Describe your procedure below.

b) Were you able to determine the speed of the object? Describe *exactly* what measurements and/or calculations you performed. (Try to be very specific here.)

c) If you perform this experiment multiple times, will the results be the same? Briefly explain why or why not.

d) Discuss with your group how you would measure the speed of an object that was not traveling with a constant speed. For example, assume your ball was placed at the top of a ramp and allowed to roll down. Explain below how you could measure the speed of the ball near the top of the ramp, somewhere in the middle of the ramp, and at the bottom of the ramp.

Congratulations, you have just defined a physical quantity (speed) by developing a procedure for measuring it. This is called an *operational definition*. Most students are surprised by the fact that they cannot measure the speed of an object directly. Speed is something that we *define* using basic measurements and calculations. It tells us how far an object has moved in a specific amount of time. For example, a truck moving with a constant speed of 60 mph would travel 60 miles each hour, 6 miles in $1/10^{th}$ of an hour (six minutes), or 88 feet every second. It is possible to use other units to measure distance and time. For example, a world-class sprinter can run 10 meters in one second. We would say that they travel at a speed of 10 *meters per second*. We will have more to say about the concept of speed in the next section.

Checkpoint Discussion: Before proceeding, discuss your ideas with your instructor!

2	*ANALYZING MOTIONS*

Our ultimate goal is to determine how the position and speed of an object change when forces act on it. But this is a more arduous task than you might imagine. In the last section, we saw that the simple act of describing an object moving with a constant speed was complicated by the fact that speed itself is a defined quantity. When an object's speed is changing, a description of the motion will be even more complicated. In fact, we may need to define additional physical quantities to describe this kind of motion.

Complicated motions are often difficult to describe using words. It is much easier to represent these motions with graphs. Fortunately, a Microcomputer-Based Laboratory (MBL) system can be used to make repeated measurements rapidly and graph them instantaneously. In sections 2.1 and 2.2, you will explore the nature of position-time graphs and velocity-time graphs. The activities in these sections will help you learn how to use an MBL system to study complex motions through the interpretation of graphs.

You may need some of the following equipment for the activities in this section:

- MBL system [2.1 - 2.3]
- Motion sensor [2.1 - 2.3]
- Low friction cart and track [2.2 - 2.3]

2.1 POSITION-TIME GRAPHS AND THE MOTION SENSOR

At this point, we have only analyzed the very simple situation of an object moving with a constant speed. We did so by using meter sticks, stop watches, tape, and some other tools to measure the position of the ball at specific times. Plotting this data on a graph allows you to view the motion of the ball from a different perspective. This graphical representation is an extremely useful way to view and analyze motions. Since we will be using it throughout the rest of this unit, it is vital that you understand the next few activities.

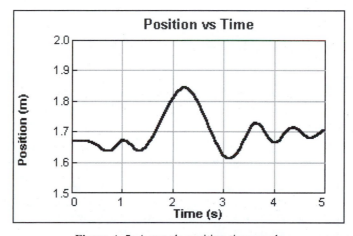

Figure A-5: A sample position-time graph

Activity 2.1.1 Constant Speed Motion

a) Without consulting your partners, make a rough sketch of what you think a position-time graph would look like for an object moving with a constant speed. **Note:** We say that the position of the object depends on the time, not vice versa. Thus, whenever we describe graphs, the dependent variable is stated first. Thus, a position-time graph refers to a graph with position plotted on the *y*-axis and time plotted on the *x*-axis.

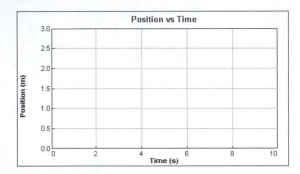

The Ultrasonic Motion Sensor

In the remaining activities in this unit, you will use an MBL system consisting of a computer, an interface box, an ultrasonic motion detector, and data collection software. It is necessary to understand some of the basic characteristics of motion sensors in order to use them effectively.

Basically, the motion sensor measures the distance of the closest object in front of it. It accomplishes this by sending out a series of sound pulses that are too high in frequency to hear (hence the term ultrasonic). These sound pulses reflect from objects near the motion sensor and some of the sound energy returns to the sensor. The computer records the time it takes for the reflected waves to return and uses the speed of sound in air to calculate how far away the closest reflecting object is. Of course, what makes the computer so useful for this task is that it can take data very quickly. There are several things to keep in mind when using a motion sensor.

1. Do not get closer than 0.5 meters from the sensor because it cannot record reflected pulses that come back too soon after they are sent.

2. The motion sensor can only detect objects within a cone about 15° surrounding the front of the detector. In addition, only the closest object to the sensor will be detected. So, be sure there is a clear path between the object whose motion you want to track and the motion sensor.

3. When measuring your own body motion, it is important to remember that loose clothing or bulky sweaters are good sound absorbers and may not be detected very well by a motion sensor.

4. The motion sensor is very sensitive and will detect slight motions. You can try to walk smoothly along the floor, but don't be surprised to see small bumps in the graph that correspond to your footsteps.

Figure A-6: A sample motion sensor. (Courtesy Vernier Software and Technology)

Position-Time Graphs of Your Motion

The purpose of the next activity is to learn how to relate graphs of your position as a function of time to the motions they represent. What does a position vs. time graph look like when an object moves with a constant speed? What does the graph look like if the object is moving at a faster, but still constant, speed? A slower but constant speed? After completing the next few activities, you should be able to look at a simple position-time graph and describe the motion of the object it represents. You should also be able to look at the motion of an object and sketch the corresponding position-time graph that represents its motion.

To do this activity and those that follow you should make sure that: (1) an interface is plugged into a power source and connected to your computer, (2) the motion sensor is plugged into the interface, and (3) that the motion software is running. You should then set up the software to record position data for a range of 0 to 3 meters for a time of about 10 seconds. Your instructor may have specific instructions on how to do this. Remember that the motion sensor detects how far away an object is. Therefore, all distances are measured as if the motion sensor is at a distance of zero (we call this the origin).

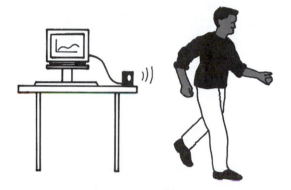

Figure A-7: Walking in front of motion sensor attached to an MBL system.

Activity 2.1.2 Trying Out the Motion Detector

a) To begin with, try something really simple. Stand one meter in front of the motion sensor and start the software. After remaining motionless for two seconds, take a step away from the sensor and again stand motionless for two seconds. Continue this procedure until the program stops running. Make a rough sketch of your graph below. Does your graph make sense?

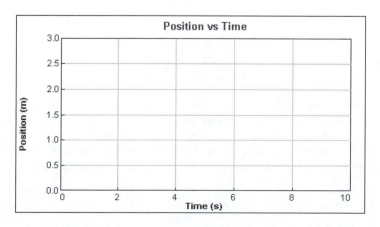

b) Once you are certain you understand how the motion sensor is working, try walking away with a constant speed (this is tougher than it sounds). Describe what your graph looks like (i.e., is it a straight line or a curve; does it slope up or down). Sketch your data on the graph below using a solid line. Then try moving away with a faster (constant) speed. Sketch the resulting data using a *dashed* line on the graph below.

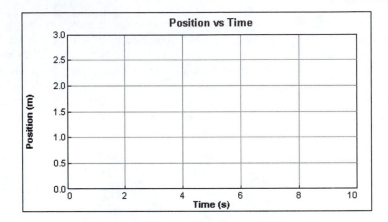

LEGEND:
Constant speed away   ——————
Greater constant speed — — — — –
Constant speed toward   ....

c) Now move towards the motion sensor with a constant speed and sketch the resulting data using a *dotted* line on the graph in part b). After some discussion with your group, write a sentence or two describing how position-time graphs look for objects moving with a constant speed either towards or away from the sensor. (If you are at all uncertain at this point, try a few more experiments before answering the question.)

Note: As simple as this last activity may seem, it is quite important that you know how to work with the motion sensor and software and also that you know what the position-time graphs look like for constant speed motion. So, if have any doubts about these topics, now is the time to experiment with the motion sensor until you have it mastered. The idea is to understand how the graphs relate to the motion, not to memorize what is happening.

Testing Your Understanding of Position-Time Graphs

If you understand what position-time graphs tell you about motion, you should be able to meet the following challenge.

Activity 2.1.3 Matching Contest!

a) A position-time graph is shown below. Each team has one shot at producing a position-time graph that looks as much like this one as possible. Get together with your group and discuss exactly how you need to move to produce such a graph. You can practice *without* turning on the computer, but remember that you only have one chance with the computer running. When you're done, print out a copy and affix it over the graph shown below. Good Luck!

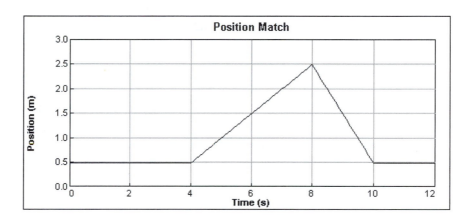

2.2 VELOCITY: A COMBINATION OF SPEED AND DIRECTION

In the last section you should have discovered that a straight line on a position-time graph represents constant speed motion. You should also have observed that the slope of the line is positive when moving away from the motion detector and negative when moving toward it. (More precisely, the direction that the motion sensor points defines what we refer to as a "positive" and "negative" direction.) The fact that the sign of the slope indicates the direction of travel is intriguing. This is our first example of obtaining physical information from a graph. It suggests that we take a closer look at the slope of a position-time graph to see if we can learn anything more about the motion of the object. Recall that the slope of a line is calculated from the expression

$$m = \frac{\text{change in } y}{\text{change in } x} = \frac{\Delta y}{\Delta x} = \frac{y_2 - y_1}{x_2 - x_1}$$

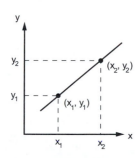

Figure A-8: The graph of a straight line.

where $x_2 > x_1$. In words, we say the slope is the "change in y (along the vertical axis) divided by the change in x (along the horizontal axis)." Now, because we are dealing with a physical situation, the slope will be seen to have a very precise physical meaning. **Note:** The purpose of the next few activities is to make a quantitative connection between the speed and direction of an object's motion and the slope of the position-time graph representing the object's motion.

Activity 2.2.1 What Does the Slope Mean?

a) Begin by making a sketch of the position-time graph for the
   following motion. An object moves with a constant speed away
   from a motion sensor. When a timer is started, the object is one
   meter in front of the motion sensor and when the timer reads 10
   seconds, the object is three meters in front of the motion sensor.
   Explain briefly how you decided what to draw.

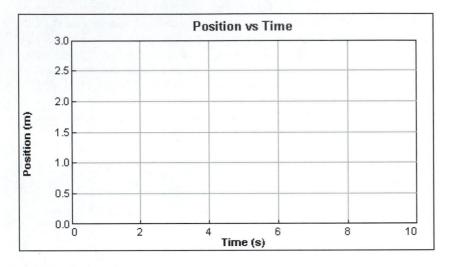

b) When calculating the slope of a line, we are free to choose any two
   points we want. Let's use the two points corresponding to $t = 0$
   seconds and $t = 5$ seconds. First, determine Δy and explain what
   this tells you physically. **Hint:** What is being plotted on the y-axis?

c) Now determine Δx and explain what this tells you physically.
   Hint: what is being plotted on the *x*-axis?

d) Now put this all together and explain what the slope of this line tells you physically.

e) Now calculate the slope of the line again (and give a physical interpretation) using the two points corresponding to $t = 0$ seconds and $t = 10$ seconds. Do you get the same value? Do you get the same interpretation? Explain briefly.

> ## Checkpoint Discussion: Before proceeding, discuss your ideas with your instructor!

It is conventional to specify the slope of a line as a single number, as opposed to a fraction. In this case, Δx is equal to one second and the slope tells you how much the position of the object changes in one second.

The Definition of Velocity

As we saw in the first section of this unit, the speed of an object cannot be measured directly. It is a defined quantity that describes how far an object travels in a specific amount of time. Notice how similar this is to the physical interpretation of the slope of the position-time graph. Recall that to determine the speed of an object you must divide the distanced the object traveled by the time it took for the object to travel that distance. But this is almost exactly how we calculate the slope of a straight line on a position-time graph: the change in position is divided by the change in time. In fact, we define the *velocity* of an object moving along a straight line with a constant speed as the slope of the position-time graph for that object.

Notice that there is an important and subtle distinction between the speed and the velocity of an object. The speed tells us how far the object travels every second whereas the velocity tells us how much the object's position changes (the final position minus the initial position) every second. This distinction is worth thinking about because it will become important later in the unit. The following activity demonstrates the difference between these two concepts.

Activity 2.2.2 Velocity and Speed

a) Consider the following position-time graph. How far has the object moved between 4 seconds and 8 seconds? What is the object's speed? How much has the object's position changed between these two times? What is the object's speed and velocity between these two times?

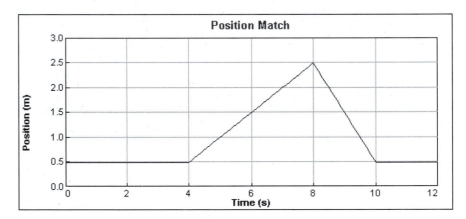

b) How far has the object moved between 8 seconds and 10 seconds? How much has the object's position changed between these two times? What is the object's speed and velocity between these two times?

c) Which contains more information, an object's speed or its velocity? Can you think of a mathematical relationship between an object's speed and its velocity? Explain briefly.

The difference between speed and velocity boils down to the difference between distance and "change in position." The distance traveled by an object is never negative, whereas its change in position can be positive or negative, depending on whether it moved in the positive direction or in the negative direction. Thus, the object's change in position takes into account which way it moved in addition to how far it traveled. Mathematically, we would write distance as the absolute value of the change in position. Similarly, the speed of an object is always positive whereas the velocity can be positive or negative, depending on the direction of travel. Therefore, an object's speed is simply the absolute value of its velocity.

Velocity-Time Graphs

In addition to viewing position-time graphs, we may also want to look at velocity-time graphs. Since the velocity is obtained by calculating the slope of a position-time graph, a computer can be programmed to find the slope, and hence the velocity, on a moment-by-moment basis. The software that runs the motion detector has this capability. The following activity is designed to help you explore the relationship between position-time graphs and velocity-time graphs. **Note:** In the next couple of activities, you will be using a motion sensor to observe the motion of a cart on a track. These carts have a small amount of friction that we would like to ignore for now. The effects of friction are more noticeable when the cart is moving slowly. Thus, whenever you are tracking the motion of a cart, make sure that it is not moving too slowly.

Activity 2.2.3 Position and Velocity Graphs

a) Set up the computer software so that it displays both a position-time graph and a velocity-time graph simultaneously. Next, set up a motion sensor to track the motion of a cart on a track. Now track the motion of the cart after giving it a brief push so that it moves freely along the track. Comment on the velocity of the cart for that portion of the motion when you are not touching the cart.

b) Use the computer software to find the slope of the position-time graph for the "freely moving" part of the motion and compare this to the velocity of the cart from the velocity graph.. (The MBL software may be capable of determining the slope of the line. Ask your instructor how to accomplish this.) Comment on your results and print out a copy of your graphs for your Activity Guide.

c) Try this experiment a few more times, altering the speed of the cart and/or its direction of travel. Again, use the computer to determine the slope and compare it to the velocity in each case. Do your results make sense? Explain briefly.

d) A velocity-time graph is shown below. Sketch your prediction for the position-time graph that corresponds to this velocity-time graph. Are there other position-time graphs that would give the same velocity-time graph? Explain briefly. **Hint:** Is there any way to tell where the object began at time $t = 0$?

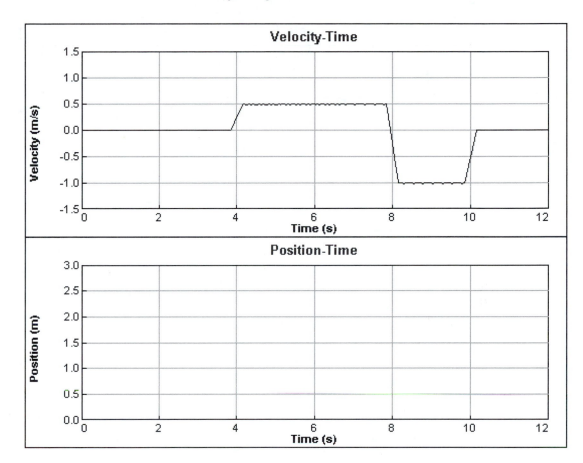

e) Now use the cart or your body to try to match the following velocity-time graph. When you have made a good matching graph, print out a copy for your Activity Guide.

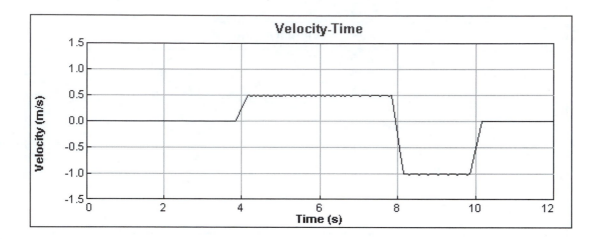

g) Two possible position-time graphs are shown below. Which graph or graphs corresponds to the velocity-time graph that you just matched? Neither? Both? The top one? The bottom one? Explain briefly.

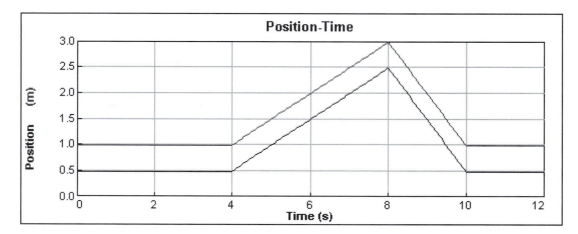

2.3   DEVELOPING A HYPOTHESIS: WHAT IS THE EFFECT OF A PUSH?

So far, we have been looking at situations where an object has been given a brief *push* to get it moving, but no other pushes or pulls are applied after releasing it. The constant speed motion that we have observed occurs after letting go of the cart. Once the cart has left our hands it is no longer subject to any direct push from us. In fact, at this point, we might even hypothesize a *law* or *rule* of nature. "An object that is not being directly pushed or pulled by anything moves with a constant speed." Of course, we don't have to look very hard to find a situation where this rule seems to fail. For example, if we push a book on a table and then let it go we would observe that the book slows down and comes to rest rather quickly. It does *not* move with a constant speed. Does that mean we need to throw out our rule? Not necessarily. Our observations so far have been with *low friction carts*, and a book on a table is not a low friction situation. In fact, it may be reasonable to suggest that the table pushes on the book (try running your hand along a table and you will surely feel the table pushing back against your hand). Therefore, we might modify our hypothesis slightly and say "in low friction situations, when there are no pushes and pulls acting on an object, the object will move with a constant speed." But what about when we are pushing and pulling on something? What then?

The Effect of a Push

At this point, we are ready to expand our investigation to include what happens when we actually push or pull on an object. This is the focus of the next two activities.

Activity 2.3.1 Observing the Effect of a Push

Figure A-9: Hand pushing cart

a)   Set up the motion detector to record the motion of a cart on a track. Start the cart moving away from the motion sensor and, when the cart gets near the end of the track, use your hand to give it a brief, firm push, sending it back towards the motion sensor. (To reduce the effects of friction, make sure the cart is not moving too slowly!) Sketch or print out the position-time and velocity-time graphs and circle the portion of the graphs when your hand is actually touching the cart

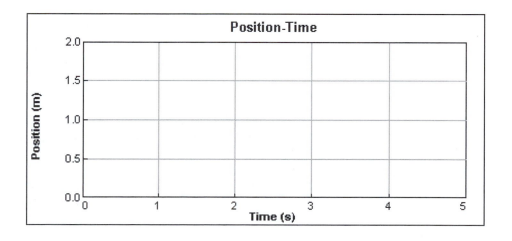

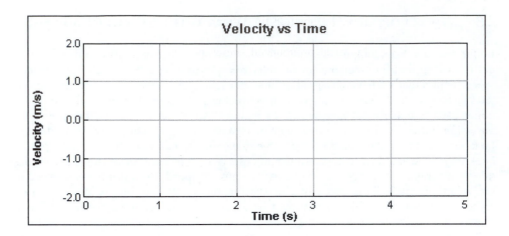

b) Explain how you can tell exactly where the push starts and ends.

c) Describe how the velocity-time graph is similar before and after you pushed on the cart. (There may be differences as well, but focus on the most pronounced similarity.)

d) Explain how the velocity-time graph looks different when you are pushing on the cart compared to when you aren't touching it.

e) Describe how the position-time graph is similar before and after you pushed on the cart. (There may be differences as well, but focus on the most pronounced similarity.) Also, explain how the position-time graph looks different when you are actually pushing on the cart.

It should be clear that both the position and velocity graphs behave differently when there is a push acting compared to when there is none. If you have extra time, try experimenting with different ways of applying pushes and pulls to the cart and check to see if the position and velocity graphs always behave the same way.

Note: As you may have noticed, when the carts move very slowly, they are not observed to move with a perfectly constant speed. This is because there is a small amount of friction acting (they are *low*-friction carts, not *no*-friction carts!). As already mentioned, the faster the cart moves, the less these effects come into play. We will come back to discuss friction later in the unit. For now, assume (pretend) that no friction acts on the cart.

What Kind of a Push Steadily Increases the Velocity?

So far, you have created a qualitative hypothesis and made a number of observations. It is time for you to refine your hypothesis based on these observations. The following activity should be completed on your own, without consulting any of your partners.

Activity 2.3.2 Creating a Hypothesis

a) Examine the graph below and describe how the velocity is changing in time.

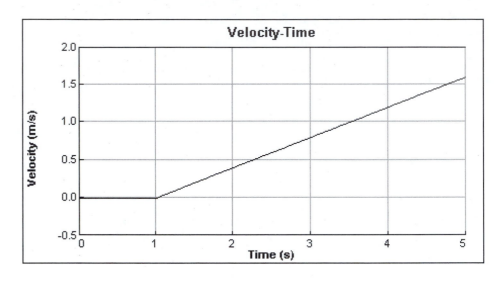

b) Explain as clearly as you can what kind of "push" you think would result in the velocity-time graph from part a). Should it be a short brief push or a long continual push? Should it get harder or softer? Should it ever be zero? etc.

What kind of pushing results in a steadily increasing velocity graph?

c) Sketch below what you think the "push-time" graph should look like to produce the velocity-time graph in part a).

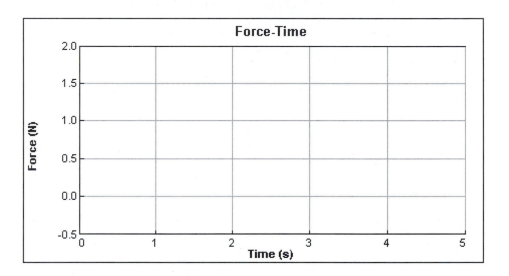

Congratulations! You have just made a hypothesis about the kind of push needed to increase the velocity of an object in a steady manner. We will be testing this hypothesis experimentally in the next section.

Checkpoint Discussion: Before proceeding, discuss your ideas with your instructor.

3	FORCE AND MOTION

We ended the last section by making a prediction about what kind of pushing causes a steady increase in speed. Unfortunately, we are not in a position to test this prediction experimentally just yet. The problem is that we have not explicitly defined what we mean by the term *push*. While it may be easy to distinguish the difference between a *hard push* and a *weak push*, these terms are vague and won't allow us to draw any quantitative conclusions. We need to be able to determine when a push is precisely twice as hard as another, or when it is only 1.2 times as hard. This requires a scientific definition of the term *push*, something that we will call *force*.

Please keep in mind that force is a loaded word, with at least 10 different definitions in the dictionary. People commonly use the word force to mean strength, energy, vigor, power, intensity, physical strength, moral strength, power to control, military power, an organized group of people (e.g. sales force), a spiritual influence, or the mystical power in Star Wars. Clearly, most of these do not apply to our investigation! It is important, therefore, not to confuse our scientific usage of the word force with these other meanings.

You may need some of the following equipment for the activities in this section:

- Rubber bands [3.1]
- Bungee cords or surgical tubes, approx. 2 ft long [3.1]
- Kinesthetic cart or other large, wheeled "vehicle" [3.1]
- Fan attachment for cart [3.1]
- MBL system [3.1 - 3.3]
- Motion sensor [3.1 - 3.3]
- Spring scale or force sensor [3.3]
- Low friction cart and track [3.3]

3.1   THE EFFECTS OF FORCE

As already mentioned, it is relatively easy to distinguish between a hard and a soft push, but we need to be able to quantify our pushes. We can do this by designing a measurement procedure for *force* that quantifies the amount of push or pull. Although there are a number of ways to define force, we will only consider one. Recall that when we began to look at motions we chose to look at the simplest case—objects moving with a constant speed. Similarly, with force, we will begin with a very simple situation; developing a method to keep the force on an object constant.

Activity 3.1.1 Defining Force

a) Hook one end of a rubber band to an immovable object. Gently stretch the rubber band a few centimeters by pulling with your finger and pay attention to what you are feeling. Can you change the length of the rubber band without changing how hard you pull on the rubber band?

Rubber band

b)   Explain how you might use a rubber band to pull on an object so that the amount of force exerted on the object doesn't change. (Your method should not depend exclusively on what you feel, since this is very unreliable.)

c)   Discuss with your group how you might define a unit of force using a rubber band (call it a rubber-band unit or RBU for short). That is, finish the following sentence, "One rubber-band unit is…" (Your definition should be very specific.) Do you think everyone else in class will come up with exactly the same definition?

d)   Now that you have defined a rubber-band unit, it is relatively straightforward to apply two, three, or 15 RBUs to an object. We simply need to use two, three, or 15 rubber bands under the conditions of your definition above. (Note that simply stretching one rubber band by two, three, or 15 times as much won't quite work because the rubber band stops stretching at some point.) Determine how many RBUs it takes to lift a heavy object, such as a bowling ball, and compare your value with another group's value. Record your results below and use this information to determine how many of your RBUs are equivalent to one of their RBUs.

Converting Between Units

As the last activity just demonstrated, it is not difficult to apply a constant pull using a rubber band nor is it difficult to determine how to apply pulls that are two or three times as large. However, exactly what we agree to call "one RBU" is completely arbitrary, in the same sense that defining a length called an inch or a centimeter is arbitrary. The important thing is that we can convert from one to the other. Converting from one set of units to another involves measuring the same item with both sets of units and comparing your measurements. This is what you did in the last part of the previous activity. This process leads to a *conversion factor*, such as there are 2.54 centimeters in one inch.

When it comes to forces (pushes and pulls), there are a few different definitions that are in wide use. In the United States, the most common unit of force is the pound. In the rest of the world and among scientists in the United States, it is the Newton. A Newton is quite a bit smaller than a pound. It is approximately the force needed to lift a small apple. For most of what follows, we will use a mechanical spring scale or an electronic force sensor that is adjusted to measure force in Newtons.

One final comment regarding forces. You may have realized that you can use your rubber band to apply a force to an object in any direction you want. You accomplish this simply by pulling the rubber band in a different direction. Thus, force is defined by a *direction* and a *magnitude*. This is similar to the concept of velocity, which has a direction associated with it in addition to the speed of the object. Therefore, like velocity, we can have positive and negative forces. A positive force is one that pulls or pushes on the object in the positive direction as defined by the motion sensor. A negative one pulls or pushes the object in the opposite direction.

Motion Resulting from a Constant Force

We can now study what happens when an object, such a person riding on a cart, is pulled along a straight line with a constant force. The velocity can be monitored quantitatively with a motion detector. To maintain a constant force, you can use a bungee cord or rubber tube stretched to maintain a fixed length. This is the topic of the next activity, which may be done as a class demonstration

Figure A-10: A person applying a constant pulling force to someone riding on a low friction cart.

Activity 3.1.2 Constant Force Motion I

a) Have one of your lab partners sit on a skateboard or other low-friction wheeled vehicle. Use a bungee cord or large rubber tube to pull on your partner with a constant force. Pay close attention to the bungee cord as you pull, making certain to keep the length constant throughout the motion. This is much more difficult than it sounds and may take some practice. When you have a clean run, print out a copy of your velocity-time graph or make a rough sketch below. Comment on what you observe.

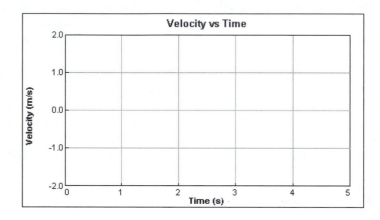

b) Make a rough sketch of the force-time graph below. This should be easy if the force was held constant!

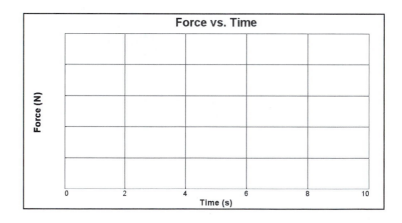

What happens when I pull someone on a cart with a constant force?

c)   Does this confirm or disprove your hypothesis from Activity 2.3.2? Explain briefly.

d)   Explain in words what effect a constant force has when applied to an object in a low friction situation.

Refining Your Experimental Techniques

Based on these observations, you can now make educated guesses as to how your partner would move under different circumstances. More observations will allow you to make increasingly sophisticated guesses, until finally the guess is no longer really a guess, but a hypothesis in which you have some confidence. The transition from a guessed prediction to a hypothesis is not well defined, and you might want to think of them in betting terms. You probably would not want to bet a large sum of money on your initial guesses. When you have enough confidence to wager money on the outcome, that is a good sign that your guess has evolved. Experimental confirmation of your predictions further enhances its validity, until it is considered a *theory* or *law* (remember that these two words are used interchangeably by scientists).

Besides making more and more observations, you can also improve your ability to devise successful hypotheses by making more accurate observations. While the previous activity does give some information regarding the motion of an object subjected to a constant force, it was also prone to all kinds of errors. For example, it is difficult to really hold the bungee cord at a fixed length when you have to move. Therefore, we will repeat the above experiment using a fan attachment on a small, low-friction cart. These more refined tools should give much more accurate results that will aid us in understanding how forces affect the motion of an object.

Using the Fan Attachment

The following activity will use a low-friction cart on a track, a motion sensor, and a fan attachment. The fan will provide a nice, constant force on the cart, but there are a few things to keep in mind when using the fan attachment.

1. Be very careful when using the fan attachments, the blades can hurt if they hit your hands.

2. Attach the fan to the cart with rubber bands so that the fan won't accidentally fall off the cart and onto the floor.

3. Always catch the cart at the end of the track before it crashes. Be gentle when catching the cart so that the fan does not fall off the cart.

4. To keep the motion sensor from collecting bad data from the rotation of the propeller, be sure that the fan blade does not extend beyond the end of the cart.

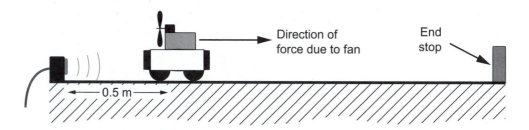

Figure A-11: Diagram of cart, fan assembly, and motion detector. Note that the fan blade does not stick out beyond the end to the cart.

Activity 3.1.3 Constant Force Motion II

a) Using words, describe as accurately as you can what you think the motion of a cart will be under the influence of the fan's steady push.

b) Make a sketch of what you think the velocity-time graph will look like for a cart feeling the steady push of a fan. Include any comments you feel are necessary.

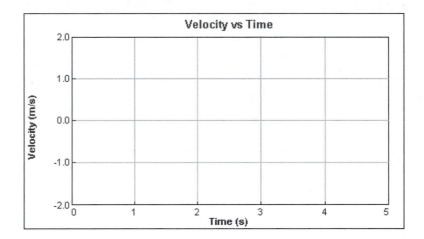

c) Now try the experiment. Using a motion sensor, produce a velocity-time graph for a cart feeling the steady push of a fan. Make a sketch of your velocity-time graph below, highlighting the portion that corresponds to constant force motion. Was your prediction correct?

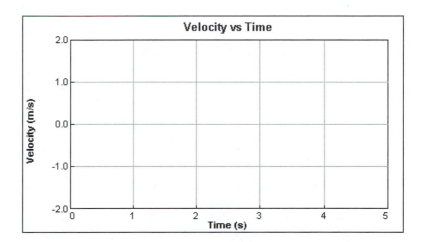

   d) Do your results support or contradict your hypothesis from Activity 2.3.2? How do they compare to Activity 3.1.2? Explain.

   e) How do you think these results would change if the steady push of the fan were larger or smaller? Try to be quantitative in your prediction. That is, what do you predict would happen to the velocity-time graph if the force was exactly twice as large?

Checkpoint Discussion: Before proceeding, discuss your ideas with your instructor.

Summary of Constant Force Motion

Notice how straight the line on the velocity-time graph is when the cart is subjected to the steady push of the fan. (The fact that the fan provides a steady push, even while the cart is moving, is something we have not verified. If this at all bothers you, talk to your instructor. Verification of this would make a nice project.) Recall that when we initially began our study of motion, we noticed that a straight line on a position-time graph represented an object moving with a constant speed. The slope of this line told us how much the object's position changed in one second, which became our definition for velocity. Now we have come across a situation that is slightly different, but also quite similar. This time, it is not the position-time graph that is a straight line, but the *velocity-time* graph. Perhaps we can learn something by investigating the meaning of this slope.

3.2   ACCELERATION

The term acceleration is commonly used when an object is speeding up. Most of you are probably familiar with stepping on the gas pedal in a car (sometimes called an accelerator) to speed up. But if we want to compare how different objects accelerate under different conditions, it is important to have a very precise meaning for the term acceleration. Thus, we will be *defining* acceleration in a manner similar to the way we defined velocity earlier. A word of caution is in order here. Most of us are familiar with the concept of speed, but the precise meaning of velocity, which includes direction, takes a little getting used to. In an analogous manner, the precise meaning of acceleration will probably not correspond to what you have always thought of as "acceleration." Like the concept of velocity, the scientific definition of acceleration will also take some getting used to.

Recall back in Activity 2.2.1, we found that the slope of a position-time graph had a very nice physical interpretation. Since we have just seen two examples where an object acted on by a constant force produces a straight line on a velocity-time graph, you might be wondering if the slope of the velocity-time graph has a nice physical interpretation as well. As you will see in the following activity, it does!

Activity 3.2.1 What Does the Slope Mean?

a)   Begin by making a sketch of the velocity-time graph for the following motion. A cart with an attached fan is placed on a frictionless track and the fan is turned on. When a timer is started, the object has a velocity of 1 m/s and when the timer reads 10 seconds, the object has a velocity of 5 m/s.

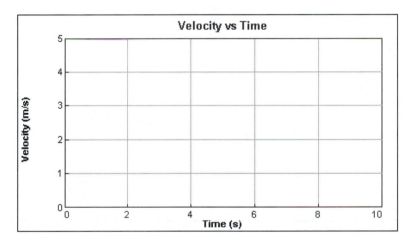

b)   When calculating the slope of a line, we are free to choose any two points we want. Let's use the two points corresponding to $t = 0$ seconds and $t = 5$ seconds. First, determine Δy and explain what this tells you *physically*. **Hint:** What is being plotted on the y-axis?

c) Now determine Δx and explain what this tells you physically.
    Hint: What is being plotted on the x-axis?

d) Now put this all together and explain what the slope of this line tells you physically.

e) Now calculate the slope of the line again (and give a physical interpretation) using the two points corresponding to $t = 0$ seconds and $t = 10$ seconds. Do you get the same value? Do you get the same interpretation? Explain briefly.

The Definition of Acceleration

This activity should have seemed somewhat familiar. If you haven't already done so, it is worth reviewing Activity 2.2.1 at this time. The mechanics of calculating the slope of the velocity-time graph are exactly the same as for a position-time graph, however, the interpretation is very different. We defined velocity as the slope of a position-time graph and it tells us how much an object's position changes in one second. Similarly, we *define* acceleration as the slope of a velocity-time graph and it tells us how much an object's *velocity* changes in one second.

Velocity and Acceleration are Not the Same Thing!

The definitions for velocity and acceleration that we have given are very similar and are easy to get confused with each other. It is essential that you remember that velocity tells you how much the object's *position* changes in one second and acceleration tells you how much the object's *velocity* changes in one second. To determine an object's velocity requires that we calculate the slope of the line on a position-time graph. Similarly, to determine the acceleration requires that we calculate the slope of the line on a velocity-time graph. In words, an object moving with a constant velocity of 10 m/s will change its position by 10 meters every second. Assuming the object begins at a position of zero meters, it will be at 10 meters after one second, 20 meters after two seconds and 100 meters after 10 seconds. Similarly, an object with a constant *acceleration* of 5 m/s^2 will change its *velocity* by 5 m/s every second. Assuming that it is not moving initially, it will be traveling at 5 m/s after one second, 10 m/s after two seconds and 50 m/s after 10 seconds. **Note:** This means that the objects position changes by a larger and larger amount each second.

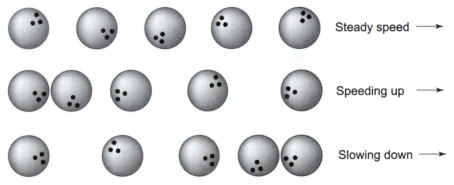

Figure A-12: Snapshots of a bowling ball taken at equal time intervals in different situations.

Activity 3.2.2 The Sign of Acceleration

a) Recall that velocity can be positive or negative, which tells us the direction of motion. Similarly, an object's acceleration can also be positive or negative, depending on whether the change in velocity is positive or negative. Make a sketch of a velocity-time graph for an object that begins moving at 10 m/s and undergoes a constant acceleration of -2 m/s^2 for 10 seconds.

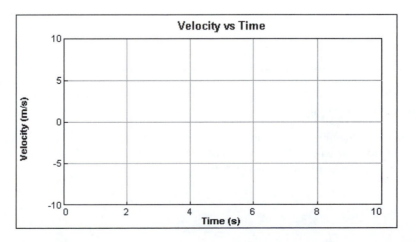

b) What are the speed, direction, and velocity of this object at 2 seconds, 5 seconds, and 8 seconds?

c) Describe the motion of this object by giving the direction of motion and stating whether the object is speeding up or slowing down. In particular, are there any times during which the acceleration is negative but the object is speeding up?

d) Is it possible for an object to have a positive acceleration while slowing down? If so, sketch a velocity-time sketch for this situation. If not, explain why not.

Putting it all Together

At this point, you may be a little confused about how position, velocity, and acceleration are all related to each other. The following activity should help you gain familiarity with these relationships by considering two very important cases.

Activity 3.2.3 Position, Velocity, and Acceleration

a) Make a sketch of the position, velocity, and acceleration graphs for an object moving with a constant velocity of +0.5 m/s, assuming the object begins at the origin. **Hint:** Begin with the velocity graph.

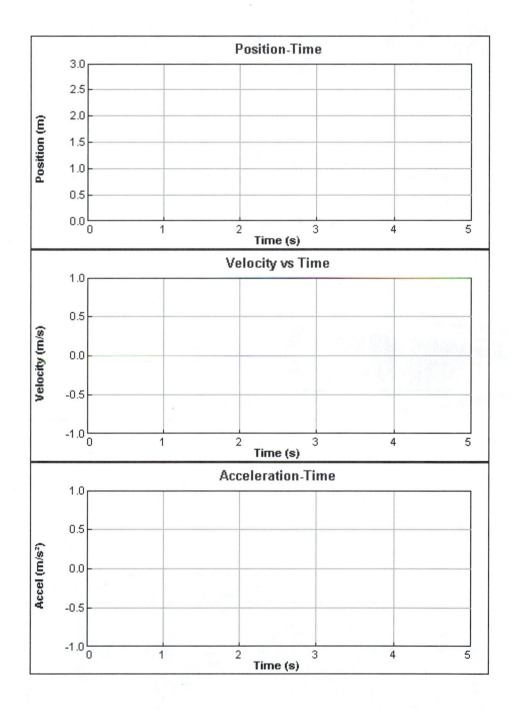

b) Make a sketch of the position, velocity, and acceleration graphs for an object moving with a constant velocity of -0.5 m/s, assuming the object begins 3 meters in front of the motion sensor.

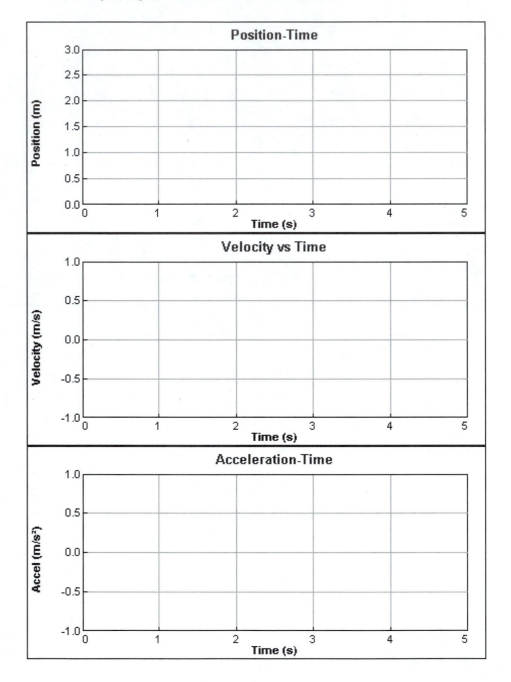

c)   Make a sketch of the position, velocity, and acceleration graphs for an object moving with a constant acceleration of +0.25 m/s^2, assuming the object begins at the origin and is not moving initially. **Hint:** Begin with the acceleration graph. Your position-time graph does not need to be very accurate, we are just interested in the shape of the graph. If you have difficulties, remember that if the velocity is changing, the slope of the position-time graph must be changing! That means it cannot be a straight line.

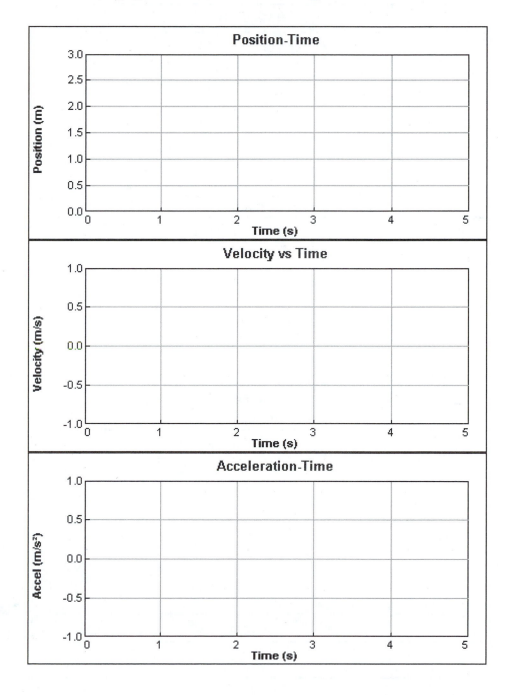

d) Make a sketch of the position, velocity, and acceleration graphs for an object moving with a constant acceleration of -0.25 m/s^2, assuming the object begins 2 meters in front of the motion sensor and is not moving initially. **Hint:** Begin with the acceleration graph. If you have difficulties with the position graph, remember that if the velocity is changing, the slope of the position-time graph must be changing! That means it cannot be a straight line.

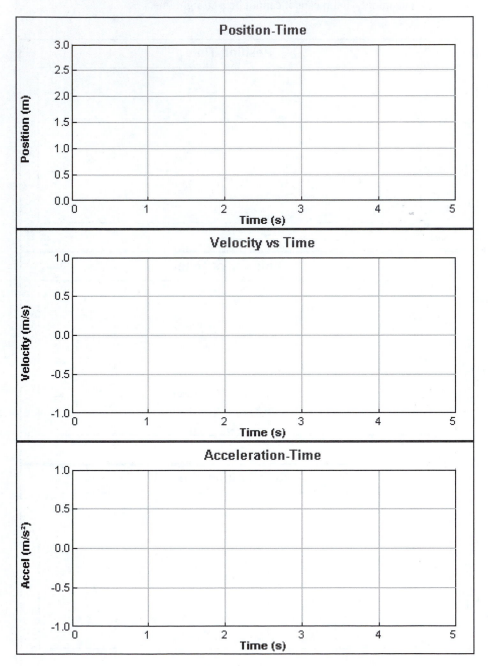

3.3   FORCE AND MOTION: THE VERDICT

We are finally ready to carefully answer the question posed at the very beginning of this module: How does an object move under the influence of a constant force? It is not enough to simply say, "it accelerates," although that in itself is a profound answer that was not appreciated until the 18th century. In the same way that we were able to quantify the velocity and determine when it was constant and when it was changing, you also know how to quantify an object's acceleration. This will allow us to determine whether an object's velocity is changing at a steady rate (constant acceleration), and if so, precisely what that value is.

Setting up a Ramp and Measuring Force

In the activities that follow, you will analyze the motion of a cart rolling down a ramp. To set this up, place one or more books underneath one end of a motion track to form a ramp. The ramp should be high enough for the cart to easily roll down. Now put the motion sensor at the top of the ramp pointing downwards. Before starting this activity, check with your instructor to see if there are any additional instructions on how to set up this experiment. **Note:** The computer will record nonsensical data (especially velocity and acceleration data) when the motion sensor is not properly tracking the cart. This most often happens when the cart is within 0.5 meters of the sensor or when the motion sensor is not properly aligned!

Activity 3.3.1 Force Acting on a Cart on a Ramp

a)   Raise one end of a track by approximately 15 cm. Place the cart at the top of the ramp and use a spring scale or force sensor to measure the force (in Newtons) that is pulling the cart down the ramp and record it below. Should this be considered a positive force or a negative force? Explain why.

b)   Make a rough sketch below of what you think the velocity-time graph of the cart rolling down the ramp will look like. Will it be a straight line? Give a brief explanation for your answer.

c) Release the cart and have the computer display a velocity-time graph. Do this a few times until you get a clean graph and sketch it below.

d) Does the cart accelerate, and if so, is its acceleration constant? What evidence do you have that this is so?

e) What do you think the acceleration of the cart would be if the force on the cart were twice as large? Explain how your velocity-time graph would change. You should be as quantitative as possible.

Testing your Hypothesis

You have just made a quantitative prediction about the acceleration of the cart that can be answered unambiguously. That is, since your prediction is a number, when you measure the acceleration you will either be right or wrong. Of course, we don't expect the experiment to give you your number *exactly*, even if your prediction is correct. Even the best experiment has a certain amount of uncertainty or error that results from not being able to measure something with perfect accuracy. There is no way around this. But if your experiment is reasonably close to your prediction, then the logic you used to make your prediction gains predictive power.

Activity 3.3.2 Testing Your Prediction

a) Increase the ramp angle so that the force on the cart (as measured by the spring scale) is *approximately* twice as large as that in the previous activity. Record the force in Newtons below.

b) Create a velocity-time graph and determine the acceleration of the cart down the ramp. Does this experiment support or contradict your prediction from the last activity? Explain your answer.

What does a force vs. acceleration graph look like?

c) Now make three or four additional force and acceleration measurements and fill in the table below. Try to obtain a large range of forces with the smallest being about 0.5 Newtons (any smaller and friction begins to play a significant role). Using the computer, graph these data with acceleration along the *x*-axis and force along the *y*-axis, and comment on its shape (does it look like anything we've seen before?).

Cart	
Force (Newtons)	Acceleration (m/s^2)

d) Use the computer to fit a straight line to your data. Your line should pass through the origin, or at least come very close. What force corresponds to a point at the origin? What acceleration corresponds to this point? Why should we expect our line to go pass through the origin?

e) Assuming that the y-intercept is actually zero, write down the equation that relates *force* to *acceleration*.

Inertial Mass

You may be wondering why a graph of force vs. acceleration would look like a straight line through the origin. The answer is simply "because that's how nature behaves." It is an experimental fact that force and acceleration are *proportional* to each other. The term proportional means that the two variables are linearly related and the line goes through the origin. Hence, if the force is doubled, the acceleration is likewise doubled. But exactly how much force is required to produce a specific acceleration? This information is contained in the slope of the line. Physically, the slope tells us how much force is needed to achieve a particular acceleration. Stated another way, it is a measure of an object's *resistance* to acceleration. We call this slope the *inertial mass* (or just the *mass*) of the object. Objects with a large inertial mass have a large resistance to acceleration and are often said to have a large *inertia*. We will investigate the influence of inertial mass on motion in the next two activities.

Activity 3.3.3 Causing a Car to Accelerate

a) A friend calls you in a panic. The battery in his car is dead and he needs you to help push it so he can throw it in gear and turn over the engine. You blithely tell him that you'll be right out. Then you remember that your friend owns two vehicles—a large delivery van and a smaller sports car. Which do you hope it is? Why?

Figure A-13: Would you rather push a sports car or a larger van? Why?

b) It is often said that in outer space things are "weightless." Do you think it would be any easier to accelerate the car (or truck) in outer space compared to accelerating it on Earth? Why? (Careful, we're not trying to *lift* the vehicle, we're just trying to accelerate it. Consider whether it would be easier to push a car that was being lifted off the ground by a crane.)

c) Would you make the same choice as in part a) if you were in outer space? Explain briefly.

Force and Mass

It is pretty obvious to most people that a more massive (heavier) object will be "harder" to push. But what exactly does that mean? If the van is twice as heavy as the car does that mean it is twice as "hard" to push? The next activity will help to answer this question. In Activity 3.3.2 you made a number of force and acceleration measurements for a cart. What you will do in this exercise is explore how the acceleration change when you increase the amount of "stuff" moving down the ramp while keeping the force on the cart constant.

Activity 3.3.4 Mass as the Amount of "Stuff"

a) Look back at Activity 3.3.2. Find the largest force you measured and copy that force and the corresponding acceleration below.

Single Cart: F =   a =

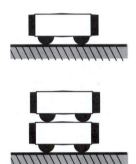

b) Now stack one cart on top of another so that the amount of "stuff" moving down the ramp is doubled. By following a procedure similar to the one you used in Activity 3.3.2, adjust the ramp height until you measure a force approximately the same as the force you reported in part a). What do you think the acceleration as a function of time will look like when the double cart rolls down the ramp? Sketch your prediction in the graph frame below. Do you think the acceleration will be a constant? If so, predict what value of the acceleration will be. Explain your answer briefly.

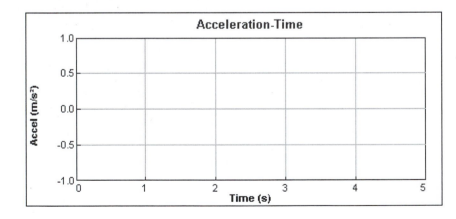

Figure A-14: How does the acceleration of the cart change when more stuff is added?

c) Now perform the experiment and create a velocity-time graph. Attach or sketch the graph below. Is the acceleration constant? If so, determine its value. Did you predict a constant acceleration. Did you predict the correct value?

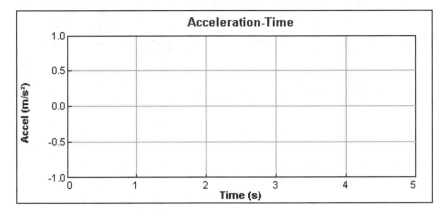

Double Cart: F =   a =

d) Repeat the above process of measuring the force and the acceleration, only this time make a stack of three carts so you have triple the amount of stuff used in Activity 3.3.2. Make sure to adjust the ramp so that the force is the same as it was in parts a) and b). What is the acceleration this time?

Triple Cart: F =   a =

e) In this experiment, the force was held approximately constant. Describe how the acceleration varied as you changed the amount of "stuff" on the cart. Try to be quantitative?

Inertia is a Property of Matter

As this activity just demonstrated, adding extra "stuff" (or *matter*) to the cart results in a smaller acceleration when the same force is applied. Thus, more matter leads to a larger resistance to acceleration. But, resistance to acceleration is what we have called mass. It seems as though there is a relationship between mass and the amount of matter contained in an object. While this may seem like an obvious statement, remember that we have defined *mass* and *matter* (the amount of "stuff") very differently.

If we are going to equate inertial mass with the "amount of matter" in an object, we must be sure that our expanded definition is consistent with our original definition of inertial mass. The following activity explores this further.

Activity 3.3.5 Does More Stuff Change the Slope?

a) Use the results from Activity 3.3.2 and Activity 3.3.4 to begin filling in the table below. Then perform the necessary experiments to complete the table. You should use a large range of forces with the smallest being about 1 Newton for the double cart system and about 1.5 Newtons for the triple cart system.

Single Cart System		Double Cart System		Triple Cart System	
Force (N)	Acceleration (m/s^2)	Acceleration (m/s^2)	Force (N)	Acceleration (m/s^2)	Force (N)

b) Have the computer draw three best-fit lines to your data and print out a copy of your graph for your activity guide. From your graph, determine the slope for the single cart, the double cart, and the triple cart. Next, use a mass balance to determine the mass of a single cart, a stack of two carts, and a stack of three carts. Do you notice any relationship between the masses and slopes for these systems? Explain.

Figure A-15: How does the amount of matter as measured by a mass balance relate to the slope of a force-acceleration graph?

c) Discuss with your group how you would use your graph to answer the following three questions: (1) If the mass stays fixed and you double the force, how does the acceleration change? (2) If the force remains fixed and you double the mass, how would the acceleration change? (3) If you doubled the mass but wanted to maintain the same acceleration, how would you need to change the force?

(1)

(2)

(3)

> ## Checkpoint Discussion: Before proceeding, discuss your ideas with your instructor.

With the slope of the force-acceleration graph defined as the mass of the object, we can write down a simple equation that encompasses all the information in the previous activity. First, we see that force and acceleration are proportionally related. That means that there is a straight-line relationship between force and acceleration, and the line goes through the origin. The fact that these lines go through the origin should not be too surprising. We have already observed that when there are no forces acting on an object then the object moves with a constant velocity. That means the acceleration is zero. Therefore, when the force is zero, the acceleration will also be zero.

Now recall that the equation of a straight line is given by $y = mx + b$. Since these lines go through the origin, they all intercept the y-axis at zero. This means $b = 0$. Furthermore, since force is plotted on the y-axis and acceleration is plotted on the x-axis, the equation $y = mx + b$ becomes $F = ma$. Here, m represents the inertial mass of the object (which also happens to be the slope of the line), which we interpret as how much the object *resists accelerating* or how much "stuff" the object contains.

Newton's Second Law

The equation $F = ma$ is called Newton's second law of motion. It describes how the motion of an object is related to the force applied to it. It says that when a net force is applied to an object, that object will accelerate (its velocity changes), and the amount of acceleration depends upon the mass of the object. For a given force, the larger the mass, the smaller the acceleration. Put another way, in order to produce the same acceleration on two objects requires a (proportionally) larger force be applied to the object with the larger mass.

Recall that earlier in this unit, we learned that an object with no pushes or pulls acting on it will move with a constant velocity (see section 2.3). It is interesting to note that this is a special case of Newton's Second Law. If there are no pushes or pulls, then the force is equal to zero. This means that the acceleration should be equal to zero. But notice that this equation says nothing about the velocity. In particular, the velocity of the object does not need to be zero! This is a very common misunderstanding. Most students believe that if there is no force acting on an object, then it must not be moving at all. If you have any doubts about this, you might want to review Activity 3.2.3.

Activity 3.3.6 Pushing a Car or a Truck

a) Let us now re-visit why it seems "harder" to push on a large truck than on a small car. Since you are the one doing the pushing, the amount of force you supply will be independent of whether you are pushing on a truck or a car. After all, you only have a limited amount of strength! Using the results of the previous activity, explain what is meant when we say it's "harder" to push the truck than push the car. **Hint:** When pushing the car or truck, the goal is to get the vehicle moving at a specific speed, say, 12 m/s. If the force is the same on the truck or the car, what will be different about their motions?

The fact that force causes motion may not be terribly surprising. The value of Newton's Second Law is in the details. A force not only causes motion, it *necessarily changes the velocity of an object*. Conversely, and quite counter-intuitively, a force is not necessary in order to maintain a constant velocity. Therefore, an object at rest will remain at rest until put into motion by a force. A cart in motion will maintain its motion unless a force is applied to speed it up or slow it down. Keep in mind, at this point we have only investigated low friction situations. We will be investigating the effects of friction soon.

| **4** | ***USING NEWTON'S LAWS TO INVESTIGATE THE REAL WORLD*** |

We have now developed a coherent theory of force and motion that applies to some specific situations (low-friction rolling carts). Now we'd like to look at some more complicated situations to see if our theory of force and motion can explain what is happening. You will see that our simple theory, derived from observing low friction carts on tracks can easily be extended to describe many other motions.

For the activities in this section, each group will need:

- Hardcover book or other rigid, flat surface [4.1]
- Small objects of different weights [4.1, 4.2]
- MBL system [4.1, 4.2]
- Motion sensor [4.1, 4.2]
- Low-friction cart and track [4.1, 4.2]
- Video analysis software and movie of tossed object [4.1]
- Small spring scale or force sensor [4.2]
- Fan attachment for cart [4.2]

4.1   UNDERSTANDING GRAVITY AS A FORCE

In this first activity, we will make some simple observations of the motion of a tossed object.

Activity 4.1.1 Tossing a Ball

a)   Toss a ball straight up a couple of times and observe its motion as it goes up and comes down. Explain how you can tell whether or not there is a force acting on the ball (be specific, you can't just say "because it looks like it"). If so, explain whether you think the force increases, decreases, or remains the same when the ball is on the way up, at the top, and on the way down.

b) Now imagine dropping two objects that have different masses (each heavy enough to be unaffected by air resistance) from the same height. Do you think the force will be the same on both objects throughout the motion? Do you think they will remain next to each other throughout the motion? Explain briefly, making certain you don't contradict yourself.

c) Place two or three small objects of different mass (i.e., pen or pencil, paperclip, coins, dice) on a hardcover book or other rigid, flat surface. Make sure the objects are heavy enough that the air will not affect them much as they fall (don't use a feather or a Ping-Pong ball). Hold the book at eye level, keeping it as level as possible. Then, as quickly as you can, move the book *downwards* and out of the way so the objects fall freely to the ground. It is essential that you do not twist the book in any way or move it upwards at all. We want the objects to fall to the ground from the same height without being pushed by the book in any way. (This may take a little practice.) Carefully watch the motion of the objects as they fall. Do they hit the ground at the same time? Are the objects next to each other the whole time? Repeat this experiment as often as needed until you are convinced of the results. Describe what you observe below.

d) Does the result of this experiment imply that the forces on the objects are the same or different? Explain carefully.

The reason we are only considering objects that aren't affected by air resistance is because we are interested in using our belief in the Law of Motion ($F = ma$) to figure out how the force of gravity behaves. It is much more difficult to try and understand the force of gravity if you look at the motion of objects that are affected by both gravity *and* air resistance. So, to make things simpler, we focus our attention on objects that are only affected by gravity. Although we won't study the effects of air resistance, it would make a great project.

The Force of Gravity on Dropped Objects

The ball you tossed at the beginning of the last activity first traveled upwards and then came back down. Since the velocity changes during this motion, we know that the ball is accelerating. From this, we can deduce that it is being acted on by a force. This is the force of gravity. We also observed in the last activity, that different objects move in *exactly* the same way as they fall. This means that different objects must undergo the same acceleration. However, if the objects have different masses and the same acceleration, then there must be different forces acting on them! This is a powerful and often misunderstood conclusion. Using our knowledge of Newton's second law of motion and simply observing that objects remain next to each other as they fall to the ground, we are able to deduce that *the force of gravity must be different on objects with different masses!*

Unfortunately, without making more quantitative observations, we are not able to conclude whether the force of gravity increases, decreases, or remains the same throughout the objects' motion. The next activity will investigate this question by using a motion sensor to track a tossed ball.

Activity 4.1.2 The Motion of a Tossed Ball

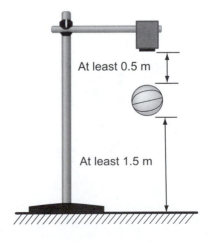

At least 0.5 m

At least 1.5 m

a) Mount a motion sensor as high as you can and orient it so that it is pointing straight down. If a basketball is dropped so that it freely falls underneath the motion sensor, predict what the velocity-time graph and the acceleration-time graph will look like *between the first and second bounce*? **Note:** Keep in mind that since you have oriented the motion sensor to face downward the positive direction on your graph will be downward.

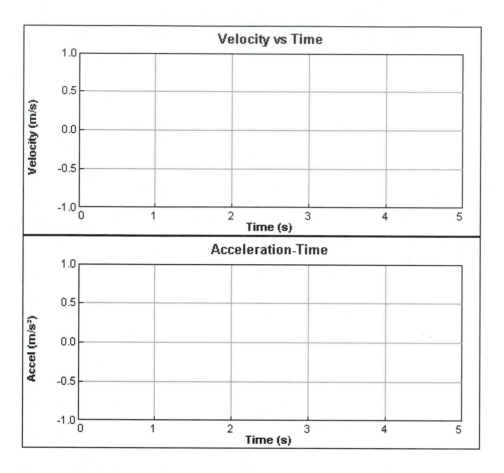

b) Now set up the MBL software so that you are viewing both a position-time graph and a velocity-time graph. Try the experiment a few times until you get a clean graph (your position-time graph should look smooth). Print out or sketch these graphs and circle the portion that corresponds to the time between the first and second bounce.

c) Is there any point in time between the first and second bounce when the velocity is zero? If so, determine the acceleration at this time. Explain your results.

d) Recall that acceleration is defined as the slope of the velocity-time graph. By looking only at your velocity-time graph, does the acceleration appear to change much between the first and second bounce?

When the velocity is zero, what does that tell us about the acceleration?

e) Try fitting a straight line to the portion of the velocity-time graph between the first and second bounce. What does this tell you about the acceleration of the ball between the first and second bounce? Use the computer to view the acceleration-time graph between the first and second bounce and compare this to your prediction.

f) What can you conclude about the force acting on the ball as it travels up and then down? Is it changing? In particular, is the force equal to zero when the ball is at the top of its trajectory?

> ## Checkpoint Activity: Before proceeding, discuss your ideas with your instructor.

Optional Activity: Using Video Analysis to Measure Motion

So far, we have been using motion sensors to track the motion of objects. Unfortunately, this is not always a convenient method of data collection. There are a number of software programs that allow one to analyze the motion of objects that have been filmed or videotape and turned into digital video clips. This provides a very powerful technique for analyzing the motion of an object. We will use one of these programs to investigate the motion of a tossed ball.

The basic idea of Video Analysis is to "point and click" on the object of interest in each frame of the movie. The movie will advance when you click on the object. It is important to click on the same part of the object in each frame. For example, when tracking the motion of a ball, it is easiest to try and place the target in the center of the ball each time.

As you click on the object in each movie frame, the computer is calculating the position of the ball within the frame of the movie. Unfortunately, the computer cannot determine the *actual* position of the object (in real life) until you tell it how. You have to provide a *scale* to allow the computer to determine what the actual distances in the movie are. Your instructor will have instructions on how to accomplish this.

Activity 4.1.3 Video Analysis of Tossed Ball

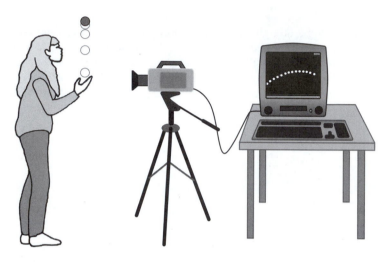

a)   Open the movie clip and play it a few times to get a feel for how the object moves. Would an analysis of the motion of this ball be applicable to all other objects thrown in the air? Explain your answer.

b)   Your instructor will have directions on how to make vertical position-time and velocity–time graphs. Comment on the shape of these graphs and sketch them below.

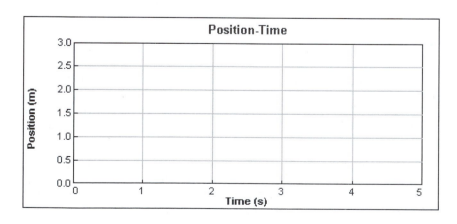

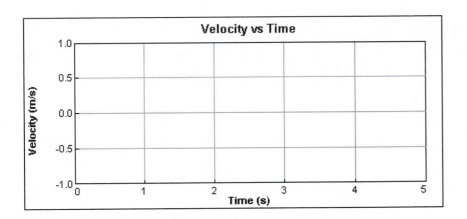

c) If appropriate, fit a straight line to the graph and print out a copy for your activity guide. Is the acceleration increasing, decreasing, or does it remain constant when the ball is on the way up, at the top, and on the way down? Provide evidence for how you know this.

d) Explain whether the force on the object is increasing, decreasing, or remaining constant when the ball is on the way up, at the top, and on the way down. Explain how you know this.

Checkpoint Discussion: Before proceeding, discuss your ideas with your instructor.

Motion Near the Earth's Surface

Many people have a strong (incorrect) belief that the force of gravity changes throughout an object's motion and that it equals zero when the object is at its peak. The evidence from the last two activities clearly supports the fact that the acceleration, and therefore the force of gravity, is constant throughout the motion of a tossed object.

We also know from a previous activity that all objects have the same acceleration when dropped. Our conclusion is that all objects not affected by air resistance experience the same acceleration of approximately 9.8 m/s² when experiencing only the force of gravity. Note that this does not imply that the force on different objects is the same. In fact, we know from $F = ma$ that if the acceleration is the same for two different objects and those objects have different masses, then the force acting on those objects must be different! The fact that the force of gravity changes in just the right way so that all objects (neglecting air resistance) fall with the same acceleration is quite profound. It was this fact that led Albert Einstein to postulate the general theory of relativity in 1916.

Figure A-16: Near the surface of the Earth, the acceleration of objects due to the force of gravity is approximately constant and equal to 9.8 m/s².

The experiments we performed to understand the force of gravity all took place near the surface of the Earth. Therefore, strictly speaking, our results are only valid near the surface of the Earth. More careful experiments have revealed that the gravitational force from the Earth gets smaller and smaller as you get farther away from the Earth. However, because of the large size of the Earth, this decrease in force occurs *very slowly*. For example, a traveler flying in an airplane 5 miles up will feel a force due to gravity that is only about 0.25% smaller than at the surface, and astronauts orbiting the Earth 100 miles above the surface experience a force of gravity that is only about 5% smaller than what they would experience on the surface.

The fact that astronauts orbiting far above the Earth have almost the same gravitational force acting on them as they do when on the surface is surprising to many people because the astronauts float as if there were no gravitational force at all. In fact, the astronauts and there spacecraft are all falling freely toward the Earth. This gives the illusion that there is no gravitational force. The reason that spacecrafts and satellites can continually fall toward the Earth without ever crashing into it is because they have a very large horizontal velocity. Their horizontal motion is so large that as they fall toward the Earth, the never get any closer because the surface of the Earth curves away at exactly the same rate that they fall. Thus, they remain at the same height above the surface even though they are continually in free-fall towards the Earth.

4.2   MULTIPLE FORCES AND FRICTION

So far, we have been focusing our attention on situations in which there is only one force acting on an object. This is why we have been using low-friction carts and using objects that are not affected much by air resistance. But now that we have a solid understanding of how one force affects the motion of an object, we can tackle the more realistic problem of how multiple forces affect the motion of an object.

Does F = ma Really Work?

At this point, our theory may seem to contradict much of your everyday experience. Pushing a safe, for example, rarely results in the safe speeding up. More often than not a constant speed is reached. Does this mean the theory we've developed only applies to special carts, and cannot be used in the real world? Well, yes and no. As it now stands, our theory rests on observations of low-friction carts, and the conclusion are strictly only valid in these situations. In fact, our results are strictly valid only in *no-friction* situations, which are impossible to produce.

As we already noted, Newton's second law has a very surprising consequence: in the absence of any applied forces an object will move at a constant velocity *forever* since there is no acceleration. This is quite contrary to many of our common experiences. Everything we observe on Earth eventually slows to a stop. As we noted before, this happens because frictional forces act opposite the direction of an object's motion. We will investigate frictional forces in the next section.

Because everything on Earth slows down, early theories for motion postulated that such slowing was the *natural* state of motion. Objects slowed because they wanted to be at rest. While this theory became complicated when applied to different motions (why, for example, do lubricants reduce the "natural" desire to be at rest), experiments to explicitly disprove the notion were impossible to obtain, since everything on Earth did eventually slow down.

The only objects that never seemed to slow down were not terrestrial, but rather up in the sky. The planets and stars were observed to move along the same path over hundreds and thousands of years. This led early scientists to believe that motion in the sky was fundamentally different than on Earth. "Heavenly" objects moved without slowing down because they were nearer to the gods, and more perfect than on Earth. Planetary motion was believed to follow the path of a circle, or a combination of circles, because the circle is the most perfect shape (being free of beginning or end).

Evidence to attack these ideas came in two forms. First, the existence of transitory objects such as supernovae and sunspots, blemishes on the "perfect" heavens, led people to believe that perhaps the heavens were not perfect, and obeyed the same laws as on Earth. Second, careful observations (most notably those by Tycho Brahe) revealed that the planets in fact did *not* move in a circular path. While other scientists ignored Brahe's data, Johannes Kepler insisted that any explanation of the heavens agree with Brahe's measurements. Using Brahe's observations as a guide, Kepler discovered that the planets moved not in a circle, but rather followed the peculiarly "imperfect" ellipse. Newton then showed how this elliptical orbit could result from a single force, and that all could be explained by relating forces to changes in velocity, rather than the maintenance of a constant velocity.

The natural state of motion, then, is constant velocity. Left truly undisturbed, any object will maintain its velocity. A moving object remains moving, an object at rest remains at rest, and an object acted on by a force will accelerate according to $F = ma$. These are the fundamental rules of motion.

Activity 4.2.1 Supporting Newton's Law

a) Provide evidence from class that supports the claim that an undisturbed object in motion will maintain that motion.

b) Provide evidence from class that supports the claim that a force causes an object to accelerate.

Multiple Forces

So far, our experiments have all been with situations where only one force was causing the object's motion. To extend our ideas to situations involving more than one force, let us begin by combining two situations we've already dealt with. Back in section 3.1, we used a fan attachment on a low-friction cart and observed that the cart's velocity increased at a steady rate. In other words, the cart had a constant acceleration. In lieu of the equation $F = ma$, that should not be too surprising since the fan supplies a constant force on the cart. We also observed, back in section 3.2, that when a cart is released on a ramp, it also undergoes a constant acceleration. Once again, the force acting on the cart must be a constant. We now want to combine these two constant forces and observe what happens.

Activity 4.2.2 Up or Down the Inclined Plane?

a) Set up a motion sensor to track the motion of a cart with a fan attachment on a level track. Orient your fan so that it provides a positive force. Set up the software so that you are viewing both position and velocity graphs simultaneously and record the motion of the cart. Be sure to turn off the fan as soon as your experiment is finished to conserve the batteries. Determine the value of the cart's acceleration.

b) Now place something small underneath one end of the ramp so that the fan is trying to push the cart up the hill. We want the cart to make it up the hill pretty easily, so don't make your ramp very steep. Again, track the motion of the cart and determine the value of the acceleration. (Turn off the fan as soon as you're done!)

c) Since the mass of the cart has not changed but you measured a smaller acceleration, what does $F = ma$ tell you about the force acting on the cart? Assuming that the force of the fan has not changed, how can you explain this result.

d) Now increase the height of the ramp quite a bit so that the fan is not able to push the cart up the ramp. Since the cart will not make it up the hill, you need to begin the experiment with the cart at the top of the ramp. Again, track the motion of the cart as it tries to make it up the hill. Determine the value of the cart's acceleration, and explain why the acceleration has changed its sign.

This activity makes it reasonable to assert that the force in the equation $F = ma$ refers not just to one force, but to the *total* force acting on the object. Although we have not demonstrated it quantitatively, the total force is obtained by simply adding together all the forces acting on the object, taking into account the direction that these forces act (some may be positive and some may be negative as defined by the motion sensor). In more general situations, forces can act in any direction, and adding them together is a bit more complicated. For simplicity, we have only been considering "forward" and "backward" forces in this unit so that their direction can be indicated by the sign of the force.

So what kind of motion results when the two opposing forces are exactly the same size? That is the topic of the next activity.

Activity 4.2.3 Equal Forces

a)   Using the same set up as in the last activity, carefully adjust the height of the ramp so that when you give the cart (with the fan turned on) a small push, it moves up the ramp with a constant velocity. You should use the motion sensor to help you determine if the cart is moving with a constant velocity. (It may be difficult to get a perfectly constant velocity, but you should be able to get pretty close. For the remainder of this activity, assume it is perfectly constant.) When you have the ramp adjusted properly, measure the velocity and acceleration of the cart. (Remember that you are to assume you've adjusted the ramp so that the velocity is perfectly constant.) What does this tell you about the total force acting on the cart? Explain briefly.

b)   Without changing the ramp, track the motion of the cart again, but this time give it a larger push up the ramp (again, with the fan on). Measure the velocity and acceleration again and explain what this tells you about the total force acting on the cart?

    c)   Finally, place the cart about in the middle of the track, but this time, don't give it any push at all (the fan, however, should be turned on). What is the velocity of the cart in this case? What is the acceleration? What does that tell you about the total force acting on the cart?

    d)   How is it possible to measure different velocities when the forces acting on the cart have not changed? Explain briefly.

> ## Checkpoint Discussion: Before proceeding, discuss your ideas with your instructor.

Summarizing Combined Force Findings

When two equal forces are acting in opposite directions on the same object, most students think that the object will not move at all. As we have just observed, that is just one of many possibilities. Most people are surprised to observe that if the object is moving, then it continues moving without changing its speed. There is a strong belief that a force is needed to keep an object moving. But consider once again what $F = ma$ says. If the total force is zero, as it is when two equal forces act in opposite directions, then the *acceleration* of the object will be zero. It says nothing about the velocity. In fact, since the acceleration is zero, that means that the velocity does not change. So if the object is not moving initially, it will continue to not move. On the other hand, if the object *is* moving initially, then it will continue to move with a constant velocity. One of the reasons why this is difficult to understand is because all of our experiences regarding motion in everyday life involve friction. We have been very careful to avoid situations in which friction is important, but the last few activities will focus on this common phenomenon.

Inferring the Existence of Forces: Friction as a Force

The ability to understand and characterize motion allows us to infer the existence of forces that are not always obvious. For example, if we push a book on a table and then let it go we observe that it slows down and comes to rest. Since the book slows down, its velocity is changing, and therefore the book is accelerating. Using our theory of motion, this implies that there must be a force acting on the book. This is a force we call *friction*. By observing the motion of the book, we are compelled to infer the force of friction.

Activity 4.2.4 Sliding Motion

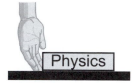

a)   Take a book, a small box, or a cart with a friction attachment and give it a quick push and let it go. Notice that it doesn't move very far before coming to rest. Now use the motion sensor to track the motion of the object while it slows down (when you are no longer touching it). Set up the computer to take data at a rate of 20 points per second and make a velocity- time graph of the motion. Print out a copy of your graphs, and circle the portion where you are no longer touching the object. Is the acceleration of the object approximately constant or not? Explain briefly how you can tell.

b)   Does the frictional force change or is it constant? Explain how you know this.

c)   In what direction does the frictional force act? Discuss this with your group and decide if it depends on the object, your push, the object's motion, or anything else.

Static Friction

Friction is a very common force that affects virtually everything on Earth. Unfortunately, it is also a very complicated force. When an object is sliding, the force of friction comes from the molecules in the object "scraping" against the molecules in the table. As we have just observed, this *sliding* (or *kinetic*) *friction* is reasonably constant for a particular object on a particular surface. If you change the object or the table, then the force of sliding friction will also change. Again, it is observed that this new force of sliding friction is reasonably constant. But what if the object is not sliding? What if it is just standing still? The frictional force behaves quite differently when the object is not moving. In this case, it is called static friction.

Activity 4.2.5 Further Investigations into Friction

a) Put a heavy weight on the table and attach a spring scale or force sensor to it. Lightly pull with a small, constant force so that the object does *not* move. Record below both the force you are pulling with and the acceleration of the object.

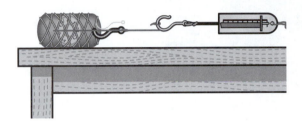

b) If Newton's Law is to be valid, what must the *total* force on the object be in order to produce the observed acceleration? What size and direction must the frictional force have to account for this total force?

c) Now pull in a different direction with a slightly larger or smaller constant force, but still small enough so that the object does not move. Again, record the force and acceleration, and explain how friction can be used to explain the observation. Has the size of the frictional force changed? What about its direction? Explain.

d) Now try pulling with a force that does not remain constant and changes direction as well (but still small enough so that the object doesn't move). Explain how the static frictional force behaves so that the acceleration of the object is always zero.

Notice how strange the *static* frictional force is. Unlike the sliding frictional force, it is completely dependent on your applied force, at least, up until you pull so hard that the object starts moving. But once the object begins moving, we are no longer dealing with static friction, but instead are dealing with sliding friction. We know from a previous activity that this force behaves quite differently than the static frictional force. We won't be studying friction in any more detail, but there are a number of aspects of the frictional force that would make a nice project.

Closing Remarks

Throughout this unit, we have developed an understanding of force and motion based almost entirely on experimental observations. We began simply enough by trying to measure the motion of an object being pushed. However, it was soon realized that in order to make any progress, we would need to be much more specific about what was actually being measured. This led us to make precise definitions for velocity, acceleration, and force. Armed with these definitions, it became a more straightforward task to determine how forces affect the motion of an object. The results of our experiments led us to conclude that a force causes an object to accelerate. This was our law of motion. The simple relationship between force and acceleration leads to the mathematical statement $F = ma$. We used this law of motion to investigate the gravitational force near the surface of the Earth, how multiple forces combine, and the behavior of static and sliding friction.

You may find it interesting to finish our study of motion by reconsidering some questions posed in the introduction of this unit. Namely, what happens if you try and push a heavy object?

Activity 4.2.6 Pushing a Heavy Object

a) If one person pushes on a heavy object, for example a safe, it is observed not to move at all. Carefully explain what forces are acting, their size and direction, and why the safe is not moving.

b) When two people push on the object, it is observed that the safe moves with a slow, constant speed. Carefully explain what forces are acting, their size and direction, and why the safe moves with a constant speed.

c) When three people push on the safe, they apply a larger force than when only two people are pushing. Carefully explain what forces are acting, their size and direction, and describe what the motion of the safe will be.

We have come a long way from the beginning of this unit. We anticipate that you now have a much stronger understanding of forces and motion than when you began this unit. More importantly, we hope that you have a much better appreciation for the process of science and the nature of scientific investigation.

5	PROJECT IDEAS

It is now time for you to take on the role of scientific investigator and to design a research project focused on some aspect of this unit that you found particularly interesting. On the pages that follow, you will find a number of project suggestions. Please do not feel limited by these suggestions. You may modify any of these or come up with a completely new one on your own. We have found that many of the best projects are those dreamt up by students. We therefore encourage you to develop your own project on a topic that you find interesting. You should of course consult with your instructor as some projects require too much time or impossibly large resources. Nevertheless, anything involving force or motion is fair game. So use your imagination and have fun!

Your instructor may ask you to write a brief proposal that outlines the goals of your project and how you plan to accomplish them. You may find it helpful to refer to the project proposal guidelines included in Appendix B. Try to plan your project in stages, so that if you run into difficulties early on you will at least be able to complete the data collection, analysis, and interpretation. To this end, it is important to note that the project proposals listed here are intended to foster your creativity, not to tell you exactly what to do. In most cases, answering all the questions in one of these proposals would take far more time than you have. So, choose a few questions that interest you or generate some of your own, but try to keep your project focused.

You will probably want to keep a lab notebook to document your project as it unfolds. Also keep in mind that you may be presenting your project to your classmates, so be prepared to discuss your results, how you measured them, and what conclusions you can draw from them. You may find it helpful to look over the oral presentations guidelines and project summary guidelines in Appendix B as you work. These guidelines may give you a better idea of what is expected from a typical student project. Be sure to consult with your instructor about their requirements for your project as they may differ from the guidelines laid out in Appendix B.

Good luck, and have fun!

5.1 DECODING FOSSIL FOOTPRINTS

© John Reader/Photo Researchers

In 1978 a team of scientists led by paleontologist Mary Leakey discovered fossilized footprints in the volcanic Northern Tanzania region of East Africa. Leakey believes these adjacent footprints had been created by two bipedal hominids that were about 4'8" and 4'0" tall. These prints provided evidence that our human ancestors were walking upright over 3.5 million years ago.

Because the hominids were walking through volcanic ash, scientists believe that their footprints were preserved through a fortunate set of circumstances. First a gentle rain turned the ash into fine-grained mud. Next, the mud was baked by the sun into hard indentations. Finally, further volcanic eruptions buried the prints in more ash that protected them from eroding over period of several million-years.

Suppose you are also a paleontologist and that you, like Mary Leaky, are searching for more tracks in an attempt to reconstruct a picture of both the physical size and the hunting practices of these early hominids. Your examination of fossil bones has revealed to you that the proportions of the limbs of these early hominids are quite similar to those of humans living today. Furthermore, you decide that its reasonable to assume that the manner in which these early hominids walk and run is similar to the way modern Homo Sapiens walk and run.

Imagine that you have been asked by the museum that your are affiliated with to answer the following questions:

1. What is the relationship between foot length and height?

2. What is the relationship between stride length and speed?

3. What kinds of factors would influence the average speed that a band of hunters could maintain when traveling all day?

4. What is the distance this individual could cover in a single day if his or her speed was given by the estimate you made above?

5.2   STREAM WATER VELOCITY

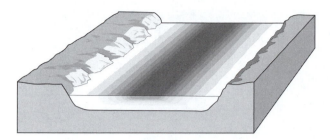

Figure A-17: A cutaway view of stream water flowing smoothly through a wide shallow channel. The darker regions in the diagram indicate areas farther from the sides and bottom of the stream.

We have become increasingly conscious of our natural environment and the need for its protection. Although it was once considered all right to use oceans and rivers as dumping grounds for waste products, we have now decided that this is an unacceptable practice. The founding of the Environmental Protection Agency (EPA) has helped to make industries more responsible in their actions. In fact, some industries are voluntarily reducing the rate of discharge of environmental pollutants into waterways.

Imagine that you work for a large corporation and are in charge of determining the maximum amount of a specific chemical that can be legally dumped into a local stream. To determine this, you first need to know exactly how much water flows in this stream. In order to get an accurate picture of how much water is transported down stream in an hour (or a day), you must find the velocity of the water at different points in the stream. For example, the speed of water ought to be different at the surface than it will be at the bottom. Why?

Once you determine the velocity profile of the stream, you can attempt to calculate the average daily flow in the stream. Your plan of action might include the following steps:

1.   Predicting the relative speed of the water flowing at key points on a cross sectional area of a stream (such as that shown along the front face of the cutaway view shown in the diagram above).

2.   Measuring the physical dimensions of a cross sectional area at a convenient location along a local stream. In other words, determining the stream's width and its depth as you move from one side to the other.

3.   Measuring the speed of the water at different depths as you move across the stream, and then devising a map or diagram that depicts the stream velocity as a function of position on the stream's cross section.

4.   Making an estimate of how much water flows past the cross section per hour, day, or year. Do you think a rainstorm will significantly effect the amount of water transported by this stream? If so, will this be a temporary or permanent change?

5.3   SHOT PUT PERFORMANCE

©Tony Duffy/Allsport   ©Bohemian Nomad Picturemakers/Corbis Images

Figure A-18: World class Shot putters Ramona Pagel and Randy Barnes must throw a heavy iron ball from within a ring that is only 7 feet in diameter

In the past few years there has been a growing recognition that scientific analysis of sports events is a key factor in helping athletes achieve optimal performance. The shot put is a prime example of an event that can be improved if the athlete fully understands how forces and motion are related.

The goal of a shot putter, like Ramona Pagel, is to heave a 10 cm. diameter iron ball weighing 4 kilograms as far as possible. After the "shot" is released it undergoes what physicists call projectile motion. A shot putter can't just throw the ball the way an outfielder in baseball does. Instead she is required to stand in a 7-foot diameter ring with the shot tucked behind her ear and then thrust the shot up and out by extending her arm as rapidly as possible. Although most of the launch speed of the shot is due to the force applied by her arm, a good shot-putter typically gains additional horizontal speed by moving across the ring while thrusting. The current world records for the horizontal distance a shot can be thrown are amazing. As of 2001 Randy Barnes still holds the men's world record at 23.13 meters (75 ft 10.25 in).

Obviously a world class shot putter must release the shot at the highest speed that is humanly possible. But other factors are important too. To study the question of how a shot putter can achieve optimal performance, you need to determine what other factors are important and how they can be optimized. If you use an experimental approach to investigating key factors you will need a way of launching an object with the same speed under different conditions. You might want to consider using digital video analysis to learn more about the nature of the resulting projectile motion. Some of the questions you should try to answer are:

1.   What factors influence the distance that a shot will travel after it is launched?

2.   What forces, if any, act on the shot after it leaves the athletes hand?

3.   Can your description of the motion be simplified by looking at horizontal motion and vertical motion separately?

5.4   UNDERWATER MOTION

©AFP/Corbis Images

Figure A-19: U.S. swimmer Megan Quann competes in the women's 4 x 100m medley relay at the 2000 Summer Olympics in Sydney, Australia. Her underwater kicking ability is crucial to maximum performance.

Around 1990, swimming the backstroke underwent a revolution due to a swimming technique known as the extended underwater dolphin kick. This technique has swimmers spending as much time under water as possible before coming up to the surface. Using this technique, the world record was shattered and it was soon clear that anyone who wanted to remain competitive would have to learn to use it.

For most of us, movement underwater does not seem as simple as movement in air. This is probably because we have much more experience with objects moving in the air. Nevertheless, there are a number of reasons that one might want to study the motion of objects underwater. As mentioned above, world class swimmers are always interested in trying to improve their times, and underwater motion was certainly the key to breaking the world record in the backstroke.

To gain an understanding of underwater motion, you should begin by designing an apparatus that allows you to drop small objects in water so that you can observe their motion. You will need to include in your plans a way of measuring velocities and/or accelerations of the objects. Some other questions you may want to explore are:

1. Does the shape of the object affect the way it falls? What about its weight?

2. What is the "best" shape for an object so that it falls in the quickest time?

3. Can you describe the object's motion using forces, acceleration, and velocity.

4. Can you describe the force from the water quantitatively? Is the force constant, or does it change? If the force changes, what factors affect it?

5.5   TERMINAL VELOCITY: PARACHUTES

©W. Wayne Lockwood, M.D./Corbis Images

We learned in class that objects near the surface of the Earth all fall with the same acceleration, regardless of their weight (or mass). This assumes that air resistance is not a major factor. However, as you are probably aware, if you stick your arm out of a car window traveling at 60 mph, you will feel quite a strong force pushing back on your arm. This suggests that air resistance can be an important factor if the speed of the object is large, regardless of how heavy the object is. In fact, a parachutist takes advantage of air resistance to dramatically slow his descent.

In order to study this affect, you will need to drop objects that are influenced by air resistance from a very high distance (as high as is reasonably possible). Dropping objects out of a third or fourth story window might be a possibility. You will also need to figure out how to measure the velocity and acceleration (if any) of the objects as they fall (you might want to consider using digital video analysis). Some of the questions to consider are:

1. Does a dropped object continue to accelerate when dropped from a very high distance?

2. Can you describe the object's motion using forces and accelerations?

3. Can you quantitatively describe the force of air? Is it constant or does it change?

4. How does the shape and/or mass of an object affect its falling motion. For example what happens to the falling motion if you create a parachute to slow an objects descent? Why?

Note: Due to the potentially hazardous effects of dropping objects from high distances, you will need to demonstrate a very well-designed plan of action and thorough responsibility for making sure no one will get hurt and nothing will get damaged!!

5.6   FACTORS AFFECTING BALL SPEED

©Clive Brunskill/Allsport   ©Ezra Shaw/Allsport

Figure A-20: Andre Agassi and Serena Williams are two of tennis' hard hitters. Is their game too fast for spectators to enjoy?

In recent years, advances in tennis racket design and player training have led to higher and higher ball speeds in professional tennis. It is not uncommon for service speeds to reach over 130 mph. While this may sound exciting, it turns out that it makes for boring tennis because there are fewer rallies to capture the spectator's interest. For this reason, the International Tennis Federation (ITF) is extremely interested in factors, such as air resistance and sliding friction, that can slow down the speed of tennis balls. A larger tennis ball can experience more air resistance. A ball with a different surface may slide differently as it bounces.

In order to slow down the game under certain circumstances, the ITF began a two-year experiment in July 2000 to allow different types of tennis ball to be used in tournaments as follows:

- New Ball Type 1 is a faster ball for use on slow surfaces such as clay. These balls will be harder and lower bouncing than standard tennis balls.

- Ball Type 2 will be used on medium paced surfaces such as hard courts and will be made to existing specifications.

- New Ball Type 3 is a slow paced ball for use on fast surfaces such as grass and some indoor carpets. Type 3 Balls will be about 8% larger in diameter than standard balls.

You've been assigned the job of determining exactly what factors play a role in sliding friction. To do so, you will need to design an apparatus that will allow you to control the force applied to type 1 and type 2 tennis balls (or other objects with different surfaces) while they slide along a court like surface. Some questions you may consider investigating are:

1.   Does the sliding frictional force depend on the speed of the ball? On the nature of the surface it slides on (e.g. clay courts or asphalt courts)? On the nature of the surface of the ball?

2.   How much will a type 1 ball slow down as it slides along a surface during a typical bounce? How much will a type 2 ball slow down under similar circumstances? Can you estimate what impact this will have on the speed of a tennis game?

5.7   STATIC FRICTION, TREAD DESIGN, AND TIRE TRACTION

Internal Design of a Tire

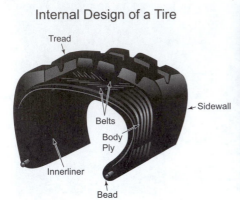

©Rick Rusing/Stone

It's a rainy day and a driver moving about 40 mph applies her brakes to slow down for a stop sign that is just around a gentle curve. Unexpectedly, she finds that her front wheels are skidding out of control on the slick road. As the driver tries to turn the steering wheel into the curve, the car continues to move in a straight line. She has lost traction.

Obviously, it is extremely important to manufacture tires that continue rolling rather than sliding on wet or icy roads when brakes are applied. A large amount of time and energy is spent trying to design new materials and come up with new tread designs to improve the static frictional force between the tire and the road. The words "TRACTION A" are shown at bottom of the Goodyear tire shown above. Traction is a rating of the tire's ability to stop on wet pavement as measured under controlled conditions on specified government test surfaces of asphalt and concrete. The traction grade (A=Best, B=Intermediate, C=Acceptable) is based upon "straight ahead" braking tests.

As an intern working for a tire company, you have been asked to make a presentation regarding traction at a special meeting of design engineers. In order to make an intelligent presentation, you decide to do some experiments to gain some first-hand knowledge regarding traction and friction. Having studied motion and friction in your science class in college, you already know a little about friction, but there are some key points that you don't understand that you want to investigate. All of these issues are centered around the question: Under what conditions does a tire lose traction? To investigate this question will require you to devise a way to measure forces while pulling or pushing on a section of tire tread across a road surface.

1. Consider a section of tire tread with a fixed amount of mass loaded on it. Is the force needed to actually start it sliding always the same?

2. How does the force needed to get a section of tire tread sliding change if the tire has a different traction rating (A, B, or C)? If the piece of tire tread is badly worn?

3. Does the load in a car or truck (i.e., the number of passenger or cargo) play a role? That is, will the same object require the same amount of force to cause it to slide if more weight is loaded on it?

4. Are there observable differences between tire treads with different traction ratings that would allow you to predict which ones will have better traction?

UNIT B
Light, Sight, and Rainbows

Detailed Contents

UNIT B

LIGHT, SIGHT, AND RAINBOWS

©Corbis Digital Stock

"...if by means those [vibrations] of unequal bigness [length] be separated from one another, the largest beget a Sensation of a Red Colour, the least or shortest a deep Violet, and the intermediate ones, intermediate colours; much after the manner that bodies, according to their several sizes, shapes and motions, excite vibrations in the Air of various bignesses, which according to those bignesses, make several Tones in Sound."

—Sir Isaac Newton

---

0   OBJECTIVES

1. To understand how we see the world around us.

2. To observe, classify, and describe various characteristics of light.

3. To use basic tools to measure properties of light such as intensity.

4. To understand how a simple lens can model the eye.

5. To use a prism and a diffraction grating to separate light into its constituent colors.

6. To learn why we see colors.

7. To analyze rainbow formation.

8. To use the concept of scattering to explain why the sky is blue and sunsets are red.

9. To learn more about the nature of light, vision, and the process of scientific research by undertaking an independent investigation.

0.1   OVERVIEW

Sight provides us with more information about the physical world than any of our other senses. Think how helpless we are in a dark room, stumbling over objects we easily avoid in the light. Yet sight is the most mysterious sense. Unlike touch, taste, or even smell and hearing, sight extends over an incredible distance. Look outside on a clear day. Your eyes easily detect the sun almost 100 million miles away! Yet you also read these words, mere inches from your face. How is sight so flexible? How does it work? How does information, whether from the sun or this page, reach your eyes? What do your eyes do with this information?

In this unit you will make observations that will help you develop a theory for vision and sight. You will discover how information reaches your eyes and how your eyes then organize this information. In the process you will also discover the role light plays in vision, how eyeglasses improve vision, what makes an object "colored," and how rainbows form.

One reason for vision's mystery is its reliance on light, an external agent. To taste an ice cream cone requires ice cream and a mouth. But to see the ice cream requires the ice cream, our eyes, and light. Why? This question has puzzled scientists and philosophers for over 2000 years. Even the seemingly simple question, "what is light," went unanswered for thousands of years. Although we will not attempt to answer this question, we can learn a great deal about vision without knowing exactly what light is. Instead, we will deal with the more straightforward question, "how does light behave?" Answering this question will lead us to investigate how light travels, how it interacts with objects, and how it enables us to see colors.

©PhotoDisc, Inc.

©Corbis Digital Stock

Section 1: How does visual information get to our eyes?

Section 2: How does the eye collect visual information for the brain?

Sections 3 & 4: How do we see color and what causes colorful visual phenomena like rainbows and sunsets?

Figure B-1: The main questions we will investigate in this unit.

Vision is a complex phenomenon. Thus, it will be useful to break our study of vision down into smaller, more manageable, pieces. Figure B-1 illustrates the three primary questions we will investigate in this unit. First, we will investigate how visual information gets to our eyes. Do our eyes "go out and get it," or does the information travel to our eyes? Once visual information gets to our eyes, it must be collected in a form that our brains can interpret. Therefore, a second valuable question is "how do our eyes collect visual information?" Our eyes collect more information than just the shape and location of objects. They also tell us about color. Thus, a third question is, "How do we see color?" None of these questions addresses the concept of perception, i.e., the interpretation of visual information by the brain. While perception is a fascinating field of psychology, it is beyond the scope of this course. We will not pursue the question of how the brain interprets information from the eye.

1	*HOW IS INFORMATION TRANSMITTED TO OUR EYES?*

Let's begin with the first of the three processes needed to understand sight by asking, "how does the information about the outside world get to our eyes?" What is light's role, if any, in this process?

Your group will need some of the following equipment for the activities in this section:

- Completely darkened room [1.1]
- Mini-Maglite™ flashlight (AA size) [1.1 - 1.3]
- Cardboard tube [1.1]
- Clear rectangular plastic container filled with water [1.1]
- Powdered creamer [1.1]
- Black Ping-Pong ball with 3-5 pinpricks and 1 larger hole to fit over Maglite bulb [1.2]
- Small block of wood [1.2]
- Ray box [1.3]
- Mirror [1.3]
- MBL system and electronic light sensor [1.3]
- Black body (light-tight box with small hole) [1.3]

1.1   LIGHT, DARK, AND SIGHT

Seeing is so much a part of our lives that most people haven't thought much about what it means to actually "see" something. The following activity poses the seemingly obvious, but intriguing question, "can we see in the dark?"

Activity 1.1.1 Can You See in the Dark?

a)   Consider the following statements by two students:

Student 1: *"Light is necessary to see. We can't see in the dark."*

Student 2: *"I disagree. By waiting for our eyes to become "dark adapted" we can see in dark rooms or outside on dark nights. Light helps us see better, but it is not necessary."*

Discuss the above opinions with your partner. Do you agree with either student? Both? Neither? What *is* necessary for us to see? Below, write your thoughts on the matter.

b) If you go into your bedroom, close the blinds (or curtains), and then turn out the lights, do you think you will be able to see? Will things look any different after a minute or so?

c) Now imagine going into the closet in your room where there is no window or light, closing the door, and stuffing rags into the cracks at the bottom of the door. Do you think you will be able to see? Will anything change after a few minutes?

d) Your instructor may be able to put you in a room that has been made "light proof." If so, briefly describe the experience (Could you see? Did things change after a minute or so?).

e) Based upon your observations, do you believe the presence of light is necessary for us to see?

Most people have never experienced complete darkness and are surprised to find out that they can't see anything at all. One fad in the late 1980's was the "sensory deprivation tank" where a person would enter a completely dark and soundproof chamber. Some people claimed that the complete lack of sight and sound led them to a "higher" state of consciousness. Others found the experience disturbing and emerged shaken. Still others were just plain bored.

So we can't see in the dark! Put another way, *the presence of light is absolutely necessary for sight*. Thus, if we are going to try and understand how we see things, it is important to understand something about how light behaves. We all know that if we look directly at a light bulb, we can see light. But, is this the only way to see light? People often talk about flashlights producing a "beam" of light. Cartoons frequently show a visible beam emanating from lights. The question we want to answer is under what circumstances does your eye detect light? If light is passing by your face, can you see it?

Activity 1.1.2 Can You *See* the Light?

a)   Do you think you can see a flashlight beam from the side? What prior experiences of yours support your idea?

b)   Look through a cardboard tube while your partner shines the light sideways past the tube, being careful not to get any light inside the tube. Can you see the beam? (Focus your attention on the *beam* itself, not on whether you see the light on the wall or on the inside of the tube.) Does this surprise you at all? Explain.

c) Below is a side view of an experiment in which a flashlight beam is pointed at a clear, rectangular container of water with a piece of paper behind it. Make a prediction as to what, if anything, you think you will see in the air between the flashlight and the container? In the water? On the paper? What, if anything, do you expect to see in these situations.

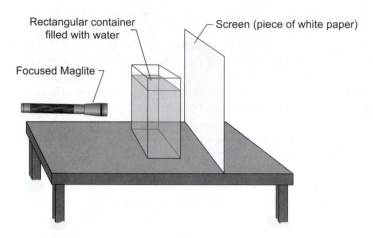

d) Now shine the flashlight through the water and onto the paper. Can you see a beam of light in the air or in the water? What do you see on the paper?

e) Do you think light from the flashlight passes through the water? Give evidence to support your answer.

f) How is the light from the flashlight involved with what you see on the paper? (Are you seeing light from the flashlight? Are you seeing the paper? Are you seeing both?) Explain as best as you can.

Developing a Hypothesis

You have just generated a hypothesis about the nature of light. A hypothesis is a conjecture (nothing more!) based upon a few general observations. Your statement about whether there is light in the water is a hypothesis, as is your idea about what you see on the paper. For the moment, you have no direct evidence supporting or contradicting your opinions. If left unsupported, a hypothesis is about as valuable as the paper on which this activity guide is printed (i.e., very little). The value of a hypothesis is that it is an idea about the world that can be tested. Everyone can generate opinions. The difficult task is developing hypotheses specific enough to be tested by experiment, or devising an experiment that tests a hypothesis. As more observations support a hypothesis it gains credibility and at some point, if it passes enough tests, the hypothesis is considered a *theory* or a *law*. We will say more about theories later.

Testing Your Hypothesis

Having developed a hypothesis, the next step is to subject it to a controlled test. That is how ideas in science are verified through observation. One aspect of science that is generally not well-understood or appreciated is that an idea can be supported with lots and lots of experimental evidence, but it can never be *proven* correct. However, it only takes one experiment to disprove an idea. This is what is known as *falsification*. A scientific idea is always open to falsification. If an idea cannot be tested and shown to be incorrect, it is not considered scientific.

Activity 1.1.3 Testing Your Hypothesis

a) Suppose you add powdered creamer to your glass of water so that it becomes slightly cloudy. Predict what will happen to the light on the paper as you add creamer. What do you think you will see when you look into the water. Explain the reasoning behind your predictions.

b) Add a small amount of powdered creamer to the water. You can always add a little more if needed. What happens to the spot of light on the paper as you make your water slightly cloudy? Does this agree with your prediction?

c) Now look into the water. What, if anything, do you see? Explain carefully what you see.

d) How does the light produced by the flashlight affect what you see in the water? How is the cream involved? (Are you seeing light from the flashlight, and if so, how is it getting in the water or to your eyes? Are you seeing cream particles? Are you seeing both?)

e) Do you think there is any connection between the dimming of the spot on the paper and the "beam" of light in the water? Explain.

Checkpoint Discussion: Before proceeding, discuss your ideas with your instructor.

In the last section, you saw that a beam of light is only visible in certain situations. In fact, there seems to be some correlation between the dimming of the spot on the paper and our ability to see the beam.[1] Light from a flashlight "lights up" whatever object the flashlight is aimed at. Since light is produced in the flashlight and the object that gets "lit up" is further away, it is natural to assume that light travels from the flashlight to the object. Experiments to confirm this theory, however, are beyond the scope of our class. Because light travels so fantastically fast (about 186,000 miles every second), it is extremely difficult to track its motion. We can, however, study some of the other characteristics of how light travels.

1.2   HOW DOES LIGHT TRAVEL?

To investigate how light travels, you will use is a special Ping-Pong ball painted black on the outside and inside with several tiny holes drilled in it. When a light source is placed inside the ball, some light escapes to the outside through the holes. Your task is to discover under what conditions you can observe this light.

Activity 1.2.1 How Can You "See the Light?"

a)   Predict what you will observe when you put a small light into the middle of a black Ping-Pong ball with small holes in it (see following sketch). Keeping your head at least 6 inches from the Ping-Pong ball, predict where you should hold your head so that you can see the small light bulb directly? (Draw a sketch if you think it is helpful.)

a)

b)   After your instructor dims the room lights, turn on your light source and slide the ball over the top so that the source is roughly in the middle of the ball. Hold a piece of white paper near the Ping-Pong ball in various places and describe your observations.

b)

Figure B-2: a) Mini-Maglite with reflector removed. b) same maglite with Ping-Pong ball (painted black inside and out)

---

[1] You may have also noticed that the spot on the paper starts to turn a bit orange after adding the creamer. We will return to this phenomenon later in the unit.

c) Place one eye at least 6 inches from the Ping-Pong ball and look at
    the light bulb through one of the holes (close the other eye). Make
    sure you are seeing the actual light bulb; it should be quite bright.
    What path do you think the light is taking between the source, the
    hole, and your eye? Draw a rough sketch of where your eye needs
    to be to see the bulb and include the path of the light from the bulb
    that reaches your eye.

Cut-away View

Ping-Pong ball
with one small
hole shown

Mini-Maglite

d) If you were to remove the Ping-Pong ball, could you see the light
    source regardless of your eye's location? What does this tell you
    about where light goes once produced in a light bulb? Discuss this
    with your group and make a formal statement about how light
    travels from a light bulb and where your eye must be in order to
    see it. Consider both the masked and unmasked cases.

Point and Extended Sources, and Light Rays

Scientists often idealize concepts to clarify their understanding. For example, the tiny light bulb of the flashlight is so small that the light it produces can be thought of as emanating from a single point. Scientists call this a *point source*. A "normal" light bulb emits light from a much larger area. This is called an *extended* source.

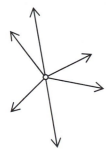

Figure B-3: A point source of light and a few light rays that are being emitted.

Before you explore the difference between extended and point sources, let's idealize further. Based on our previous observations, let's hypothesize that light travels in straight lines. To represent this, we will use *light rays*. A light ray shows a small bit of light traveling in a particular direction. Figure B-3 shows a point light source and a few (of the *many*) light rays that are being emitted by it.

Are light rays useful? In the next activity you will investigate how shadows form. By using light rays to represent light, you can make easy-to-interpret sketches of your experimental setup. Figure B-4 below demonstrates how a light ray diagram might be used to analyze a block creating a shadow on a piece of paper. It is important to always remember that we sketch only a few, of the many, light rays emanating from a source. In addition, most light ray diagram are only two-dimensional, whereas an actual point light source will also emanate light rays in three dimensions.

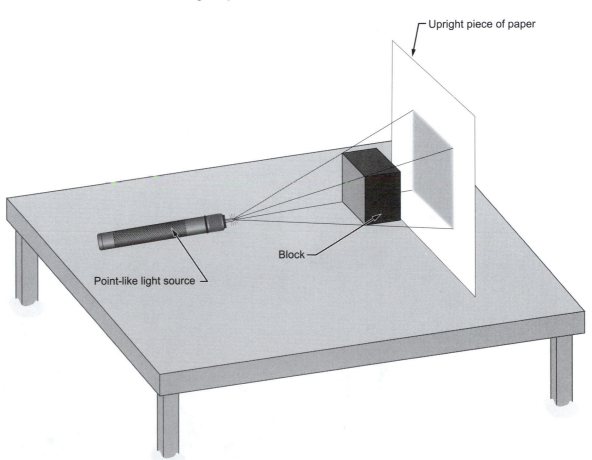

Figure B-4: Some of the light rays from the point-like light source are obstructed by the block. As a result, we see a shadow on the piece of paper.

Activity 1.2.2 Shadows—Enlightenment from Darkness

a)   A top view of a light source, a block of wood, and a piece of paper
     follows. Draw in at least six light rays emanating from the source.
     Include some that hit the block and some that don't. Show exactly
     where the block's shadow will be located.

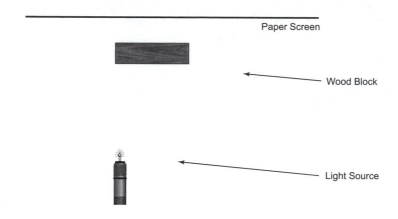

b)   Now consider what happens when there are two point sources of
     light. Again, draw about six light rays from *each* source and
     indicate where the shadow will be. Explain how this shadow is
     different from the shadow cast with only one light source?

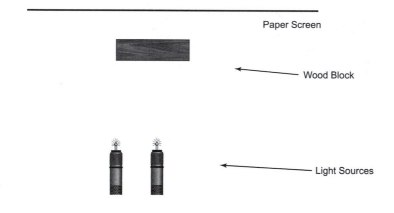

c) Set up these two experiments and describe how the shadows cast by one source and by two sources look. Are there any differences? Were your predictions from parts a) and b) correct?

d) Based on your observations, how do you think the shadow would change if there were a third light source in between the other two? If there were 100 light sources? (After answering, you might want to try it to confirm your hypothesis.)

e) Based on our earlier observations about viewing light through small holes in a Ping-Pong ball, we have been drawing light rays as if the light travels in perfectly straight lines. Is this assumption consistent with your observations from this activity? Cite evidence to support your position.

f) Your instructor may have an extended light source for you to use. If not, use the room lights to cast a shadow of the wood block on a piece of paper. Does the shadow have sharp or fuzzy edges? Explain why it appears the way it does. Use a drawing if it will help.

What Makes a Good Point Source of Light?

Many factors affect whether a light source is effectively a point source. One is the size of the light source itself. The smaller the light source, the more it will look like a point source. Another is how close you are to the source. Even a fairly large light source will look like a point source if you get far enough away from it. In general, the further away a light source is, the more it will look like a point source. A common example of an extended light source acting almost like a point source is the sun. Because it is 93 million miles away, the sun is, for many purposes, an adequate point source even though it is almost 575,000 miles in diameter. Nevertheless, there are some subtle reminders that the sun is in fact not a point source. If you go outside on a sunny day and look at your shadow, you will see that the edges are blurry. The blurry edge you see is a sign that the light source is not quite a point source. If you go back and cast a shadow with your point source, you will see a nice sharp edge to it. The more extended the light source, the less sharp are the shadows. Next time you have a chance, take a look at your shadow (if you can find it) when standing under a long fluorescent light (a *very* extended source).

1.3   LIGHT AND OBJECTS: REFLECTION AND SCATTERING

Let's take a moment to review. Our goal is to understand how we see the world around us. Our first conclusion was that light is a necessary component of vision. We therefore spent the last section learning how light travels in transparent materials, such as air or water, once it is produced in a light source. In fact, we have developed hypotheses about how one sees light from a source. We have also observed what happens when light passes through water and when it hits objects that are not transparent such as wood blocks, paper or creamer particles. Now let's extend our study of light by investigating how light interacts with a mirror.

Activity 1.3.1 Reflections on Reflection

a)   Below is a sketch of a light ray hitting a mirror at a certain angle from a line perpendicular to the surface (called the *normal* line). Draw in your prediction for how the light ray will travel from the mirror. Do you think the direction of the outgoing light ray depends on the angle with which the incoming ray hits the mirror? Explain briefly.

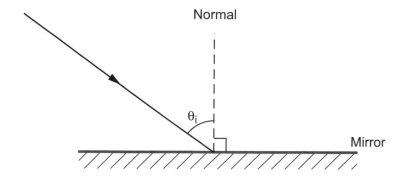

b) The figure below is designed to help familiarize you with normal lines. Normal lines have been drawn in for three of the dots already. Complete the diagram by drawing a normal line for each of the remaining dots.

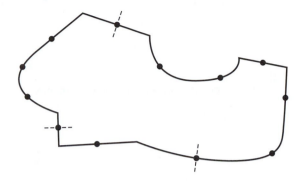

c) Using a ray box, arrange a thin beam of light so that it travels along the direction shown below. Find at least 3 different places to put a mirror so that the ray passes through the center of the dot on the right (•). Mark the location and orientation of the mirror and the path of the light ray for each case. (Use a ruler when sketching in your light rays.)

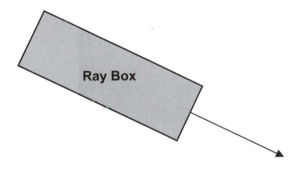

d) Are your sketches consistent with your prediction from part a)? Give a precise rule that describes how the light beam is affected by the mirror.

e)   Based on your observations, predict the orientation of a mirror that, centered on the "X", would reflect a beam of light from the ray box through the center of the dot (•). Sketch your prediction with a dashed line, and include any other lines (normal lines, etc.) you used to make your prediction.

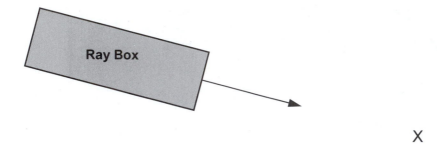

X

f)   Now try the experiment. Place a mirror, centered on the X, so that it reflects the light through the center of the dot. Draw a solid line to indicate the orientation of the mirror. Does your prediction agree with your observation?

Checkpoint Discussion: Before proceeding, discuss your ideas with your instructor.

You have just developed a *theory* of reflection from flat surfaces. A scientific theory is not just an opinion. It has been supported by observations, and makes accurate predictions about some new observations. You used your theory to predict the location of the mirror in the previous activity. This prediction also served as a test for your theory. As previously mentioned, a theory about how the world works can never be proven correct in the philosophical or logical sense. It only takes one experiment to prove that it is incorrect. Thus, we cannot say with certainty that such an experiment won't be found in the future.[2] Nevertheless, as more and more observations support a theory we begin to put more faith in its ability to describe how the world works.

---

[2] Even a technically incorrect theory can be profoundly useful. Newton's laws of motion, for example, have been superseded by Einstein's theory of relativity. These two theories differ significantly only in very extreme cases, however, and so Newton's laws are still accurate and useful in most practical applications.

Measuring Light Intensity Electronically

Because we could see the light ray on the paper in the previous activity, we didn't need any special tools to determine where the light was. This is not always the case. Although our eyes are quite sensitive to the presence of light, they are not as sensitive at detecting differences in brightness. Put another way, you can quite readily see even tiny amounts of light but it is difficult to distinguish between a little light and a little more light. A light sensor, on the other hand, is capable of measuring small differences in light intensity.

You might know that as a light gets farther from your eye it appears less bright. For example, although many stars are much brighter than our sun, they are also much farther away and thus appear less bright than the sun. This is why we don't see stars when the sun is up. Here we will investigate how lights dim as they get farther away from our eyes. In the process, we will familiarize ourselves with the light sensor.

Activity 1.3.2 Sensing Light

a) After your instructor explains how to set up and use the light sensor, use it to detect light from a flashlight. Your instructor may have to darken the room. How must the sensor be oriented with respect to the flashlight in order to detect light? Can it detect the light from the side? Is this consistent with your observations when you tried to see the flashlight with your eyes (Activity 1.1.2)? Explain briefly.

b) Now use the light sensor to detect light from a point source. What happens to the intensity as you move the sensor away from the light? Is this similar to what you observe with your eyes?

The previous activity may not seem very profound, but it is important to realize that the light sensor behaves in a similar manner to our eyes when it comes to brightness. As already mentioned, there will be situations when we cannot distinguish with our eyes whether there is more or less light in a particular experiment. In these situations, we can use the light sensor to help us.

Activity 1.3.3 Light on an Object: Part I

a) Use your ray box to produce a wide beam of light and shine it on a mirror as shown below (the dotted line shows where the center of the beam will hit the mirror). Use your law of reflection and the normal line to make a rough sketch of where the beam goes after it hits the mirror.

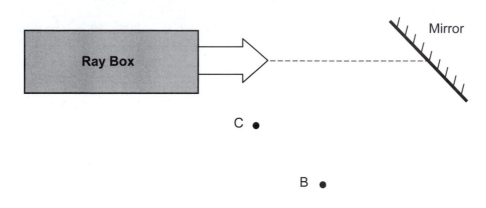

b) In the preceding diagram, you will see three points labeled A, B, and C. Place the light sensor at each of these points and aim it directly at the spot where the light is hitting the mirror. Describe the results of the light sensor. Where does the light appear to be going after it hits the mirror?

c)   Is there one point or many points from which *you* (using your eyes) can see light from the flashlight in the mirror? Is this consistent with the light sensor measurements?

d)   Now tape a white piece of paper directly to the front of the mirror, and make the same set of measurements. Is there one place or many from which the light sensor detects an appreciable amount of light? What does this imply about where the light goes once it hits the paper? **Note:** The room lights may need to be dimmed for this experiment.

e)   Is there one point or many from which *you* (using your eyes) can see light from the flashlight on the paper? Is this consistent with the light sensor measurements?

f)   What do you think the paper is doing to the light that the mirror is not doing? Why might the paper do this? Explain briefly.

There is a big difference between light hitting the paper and light hitting a mirror. In the case of a mirror, the light goes off in one direction, which depends on the angle that the incoming light makes with the mirror. With the paper, however, the light seems to bounce off in all directions. We will distinguish between these two types of behavior with the terms *reflection* and *scattering* respectively.[3]

Scattering simply means that when incoming light strikes an object, the outgoing light travels in all different directions. Reflection, on the other hand, is when most of the incoming light goes off in one particular direction. Reflection typically occurs when the surface is very smooth. A lake on a windless day acts very much like a mirror, making beautiful reflections of whatever is behind it. When the surface of the water is rough, however, light scatters in all directions and the mirror-like behavior disappears. Most objects both scatter and reflect light. What we see depends upon whether more light is scattered or reflected. This difference is very important to our ability to see objects, as you will discover in the following activities.

What are You Seeing When Looking at an Object?

So far, we have specifically avoided the most important question of this whole section. That is, "what are we seeing when we look at an object?" We know from a previous activity that without light we cannot see. We also know that in order to actually see light that is produced in a light source, our eye must be on an unobstructed line with the source. Nonetheless, we can see objects that do not produce light. The next activity addresses the question of how we see objects that reflect but don't produce light.

Activity 1.3.4 Light on an Object: Part II

a)   Below is a top-view sketch of a flashlight shining on one side of a cardboard box with rough sides. Based on your observations in the previous activity, predict what will happen to the light after it hits the box. Which of the labeled points (A, B, or C) will receive light from the box?

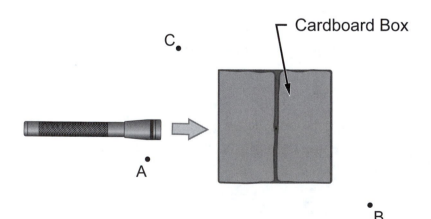

---

[3] It is also common to refer to these two phenomena as specular reflection (reflection) and diffuse reflection (scattering).

b) Now set up the above situation and use one of your eyes (keeping the other one closed) to determine whether or not light is being received by the box at each of the points A-C. Then use the light sensor to measure whether light from the box reaches points A-C. Explain both sets of observations.

c) Consider the following statements by two students:

Student 1: *"We see objects because light hits the object and then comes to our eyes. We don't see the object, we see the light."*

Student 2: *"I disagree. Once the light makes the object visible by shining on it, we can see it without any light coming to our eyes."*

Discuss the above opinions with your partner. Do you agree with either student? Write your thoughts below.

The next activity should help you resolve the students' argument in the last activity. This activity puts together much of what has been learned so far. It is recommended that you take your time and make certain you understand each question. In the next activity you will examine a "black body", which is nothing more than a closed, empty box with a small hole cut in one side. The box should be sealed so that the only way light can get in or out is through the small hole.

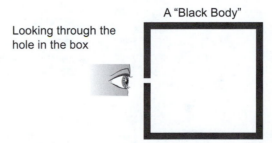

Figure B-5 : A black body is simply a sealed box with a small hole cut in the side.

Activity 1.3.5 A Black Body

a) Look carefully through the hole from different distances and describe what you see. Can you tell what color the inside of the box is? From your observations, do you think there is any light leaving the hole? Can you tell why such an object is called a "black body"?

b) Place the light sensor directly against the side of the box and take readings as you slide it back and forth over the hole a few times. Do you see any change in reading when the sensor is over the hole or not? Explain what this tells you about whether light emerges from the box or not.

c) Now imagine that there is a small light source inside the box such that you cannot see the bulb directly, as shown in the sketch below. Explain carefully what happens to light from the bulb, and what, if anything, you will see when looking into the hole. Do you think you will be able to tell what color the inside of the box is? Include a rough sketch that supports your answer.

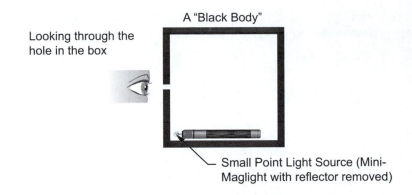

Looking through the hole in the box

A "Black Body"

Small Point Light Source (Mini-Maglight with reflector removed)

d) Now try the experiment. Fasten a small point source of light (your Mini-Maglite in "candle" mode works well) to the inside of the box and seal it up. Keeping your eye about 6-12 inches from the hole, can you tell the color of the inside of the box?

e) Finally, use the light sensor to measure whether any light is emerging from the hole by placing the light sensor directly against the side of the box and sliding it back and forth over the hole a few times. Do you see any change in the readings when you do this? What does this tell you?

f)   Must light travel to your eyes in order for you to see? Explain what
     observations support your conclusions.

Checkpoint Question: Before proceeding, discuss your ideas with your instructor.

At this point, you should have a fairly solid understanding of what it means to see an
object. If you have any questions about this, make sure you talk to you partners and your
instructor before moving on. The rest of the unit will rely on you having a solid
understanding of these concepts.

2	*FROM LIGHT TO SIGHT*

In the last section we observed different characteristics of light. We observed that light travels straight through air and water, and how light interacts with objects, reflects off of mirrors and scatters off of rougher surfaces. We also saw compelling evidence that when you see an object, light from that object is entering your eye. However, it is not obvious whether that light carries information about the object to our eyes or whether its presence is merely coincidental.

Assuming that the light is somehow conveying information to our eyes raises some questions. For example, in a well-lit room, our eyes are bombarded with light scattering from every object in the room. How do our eyes "focus" on one particular object? That is, when you look at one particular object in a room, you may see many other things out of your peripheral vision. Nevertheless, you are clearly looking at one specific object. How do our eyes accomplish this?

In this section, we will investigate what the eye might do with light that falls upon it. While eyes are often said to be the "window to the soul", they are all surprisingly similar. A picture of the front surface of an eye is shown in Figure B-6. Using basic observations on how the eye responds to changes in the surrounding environment we will construct a simple model that mimics the behavior or the eye. We will finish this section by refining this simple model and developing a more sophisticated model of the eye.

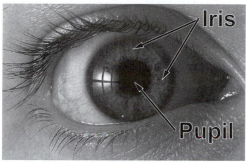

Figure B-6: A photograph of the human eye showing the iris and the pupil. (©Adam Hart-Davis/Photo Researchers)

You will need some of the following equipment for the activities in this section:

- Mini-Maglite™ flashlight (AA size) [2.1]
- Ruler [2.1]
- Bright light source (e.g. slide projector) [2.1, 2.2]
- Optical bench [2.1, 2.2]
- Variable aperture [2.1, 2.2]
- Small objects (3.5 inches) [2.1]
- Small pencil (golf pencil) [2.1]
- Black screen w/2 different colored circles [2.1]
- 4 lens/screen holders [2.1]
- White screen [2.1]
- +10 cm focal lens [2.1]
- Small Plexiglas rectangle/trapezoid with frosted face [2.2]
- Protractor [2.2]
- Ray box [2.2]
- 2-Dimensional bi-concave and bi-convex Plexiglas lenses [2.2]

2.1   MODELING THE EYE

In the following activity you will begin to observe how the eye looks and behaves. Based on these observations, you will design a physical model to try and understand how the eye works. Let's begin by exploring what function the pupil serves.

Activity 2.1.1 Examining the Outside of the Eye

a) Carefully look into one of your partner's eyes. Shine a flashlight on or near the eye and record how it responds to the increased light over a period of several seconds. What happens to the black spot found in the center of the eye (the pupil) when the light is increased or decreased?

b) Estimate the diameter of your partner's pupil in both bright and dim light. **Warning:** Do not stick anything in your partner's eye. Write down your results below.

c) Why do you think the pupil changes its size? List as many reasons as possible.

d) People's irises are different colors but the pupils are always black. Why do you think the pupil is black? What might be happening to the light that impinges on the pupil? **Hint:** Might this be related to any of the observations you made in Activity 1.3.5?

e) Do you think that the world would look different to us if our pupils were not round but slit-like, similar to a cat's pupils? Explain briefly.

The "Hole" Eye

We have had two experiences with something appearing very black. One was in the completely darkened room, in which *everything* was completely black. The other was when we looked through a small hole into a sealed box (the black body). In this case, the hole itself looked completely black. Because of these two experiences, you may have concluded in the above activity that the pupil was a small hole. Since there is no tiny light inside our eyes, we can't see inside the "box." Maybe this is why optometrists shine a bright light into patient's eyes when they look into the eye (so they can see inside). Perhaps the pupil gets larger or smaller to control the amount of light that enters the eye. This seems reasonable, so let's assume for now that the pupil is in fact such a hole and explore what happens when light from an object passes through a small hole.

In examining the behavior of a small hole and how light is affected by it, we will be drawing lots of light rays from objects. When doing so, use a ruler to help you draw straight lines.

Activity 2.1.2 The "Hole" Eye

a) The following diagram shows a pencil illuminated by a bright light. Our model of the eye is simply a small hole (representing the pupil) with a screen behind it (representing the back of the eye). Just like a piece of paper or a box, light that hits the pencil will scatter off in all directions. Draw at least five light rays that scatter from the tip of the pencil and travel towards the viewing screen including one that makes it through the small hole. Do the same for five light rays that scatter from the blunt end of the pencil. Do you think anything will be seen on the screen when the light is on? Explain briefly.

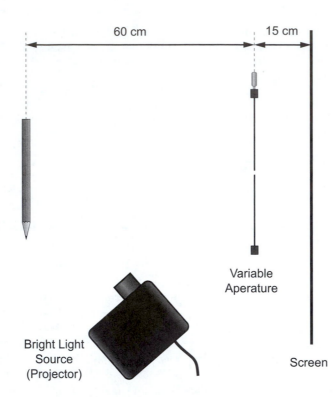

60 cm   15 cm

Variable
Aperature

Bright Light
Source
(Projector)

Screen

b) Now try the experiment. Use a very bright light, such as a slide
    projector, and an easily recognizable object (a pencil or some
    colored circles). The room light should be dimmed for this
    experiment. Describe below what you see on the screen (look
    carefully). Also describe what happens as you move the screen
    closer to and farther away from the hole. Try using your hand as
    the object.

c) The following sketch shows the same set-up, but the pencil has
    been replaced by a piece of black paper with two colored circles on
    it. Sketch in several light rays to determine where the image of the
    circles will be (you can use colored pencils if you have them).
    Note: You should draw at least four light rays; one from the top
    and bottom of each circle, that pass through the hole and hit the
    screen. Don't forget to use a ruler.

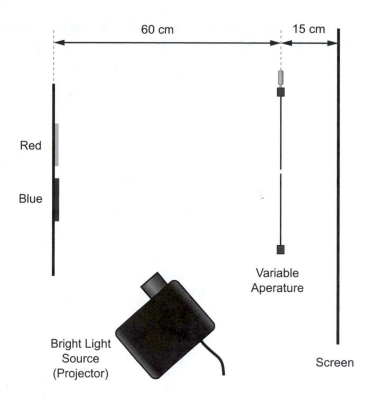

d) Are the colors on the screen in the same order (top to bottom) as they are on the card?

e) Given the distances shown in this sketch, predict the size of the images of the circles compared to the size of the original circles. **Hint:** How would the size of the image change if the viewing screen or the object was further away?

f) Now set up this experiment and test your prediction. Make the space between the aperture and the screen 15 cm and the distance between the circles and the aperture 60 cm. What size is the image compared to the object? How does this compare with your prediction? Does moving the object or the viewing screen cause the image size to change as you would expect?

Most people are quite surprised to find that a small hole can produce an image. As you have just observed, it is not necessary to have anything more than a small hole to form an image of an object on the screen. So is that all there is to our eyes? Although this simple model can indeed form an image of an object, this next activity will demonstrate some discrepancies between this crude model and our actual eyes.

Activity 2.1.3 Big Holes, Small Holes

a) Dim the light on the object slightly by placing a piece of frosted glass or waxed paper in front of the light. What happens to the image on the screen?

b) What could you do to restore the image to its original, bright, appearance? **Hint:** How does your pupil respond to dim light? Do not actually try out your ideas at this time.

c) Before performing the experiment, let's predict what will happen as our "pupil" dilates. In the sketch that follows, the hole between the dots and the screen is now much larger. This means that many more light rays will pass through the hole. To get an understanding of what effect this will have on the image, let's choose one point on each of the colored circles and look at where light from these points ends up on the viewing screen. Draw at least five light rays that scatter from the very top of the upper circle; at least three of which pass through the hole. Do the same for light scattering from the very bottom of the lower circle. **Note:** Using colored pencils will make your sketch much easier to understand.

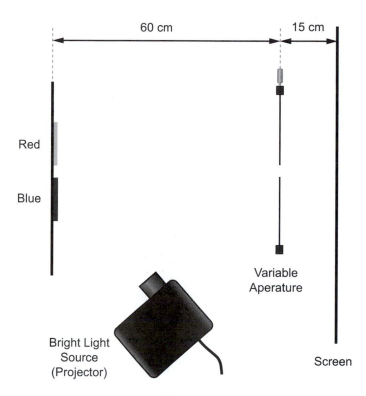

d) Examine your drawing carefully. Since the light rays you drew all come from exactly one point on each circle (which could be *any* point on the circle), do you expect to see a crisp clear image of the circles on the screen? Explain why or why not. How will things change if the pupil is made even larger?

e) Based on your drawing in part c), what do you think will happen to the brightness, size, and sharpness of the image of the red and blue circles in your model eye if you enlarge the aperture.

f) Now try the experiment. First, increase the size of the aperture to about 8 mm. This is approximately the size of a real pupil in dim light. Then, continue increasing the size of the aperture a little at a time, as you carefully watch the image. Record your observations below. Was your prediction correct?

Checkpoint Discussion: Before proceeding, discuss your ideas with your instructor.

You have observed that light that has been scattered from an object can pass through a small hole and form an image of that object on a screen. This only works, however, if the object is brightly illuminated and if the hole is very small. In dim light the image fades. Enlarging the hole to allow more light to pass through succeeds in brightening the image but results in a blurry image. As you have already observed, the hole in our real eye *does* change size, but objects don't tend to get fuzzy when this happens. Therefore, if the eye somehow forms an image for the brain to interpret, there must be some other component that helps to keep the image sharp when the pupil changes size.

Creating a More Sophisticated Model of the Eye

As you are probably well aware, our eye consists of more than just a hole. Medical scientists have dissected the eye and a picture of their findings is shown in Figure B-7. Note the pupil at the front of the eye and the retina at the back. From our previous experiments, you can see that the role of the viewing screen was to mimic the retina.

You may have noticed that one rather obvious component of the eye that we left out of our earlier model is the lens. Is it possible that the lens is responsible for creating a sharper image when the pupil is dilated? To investigate this question we will use a glass lens. Lenses are pieces of clear material with curved surfaces, and are used in eyeglasses, magnifying glasses, microscopes, and other optical instruments. In the following activity, you will observe how light behaves as it passes through a lens. Your observations should allow you to understand how the eye reorganizes the light to form a clear image, regardless of the size of the pupil.

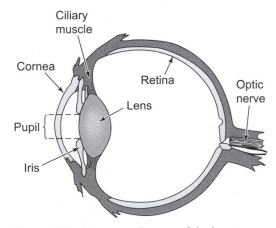

Figure B-7: A cutaway diagram of the human eye, showing some of its major components.

Activity 2.1.4 Observing a Lens in Action

a) Set up the following arrangement with the aperture opened to the size of the pupil in dim light (about 8mm). Place the lens between the aperture and the screen and slide it back and forth. Describe what you see. Can you obtain a clear image with the large aperture?

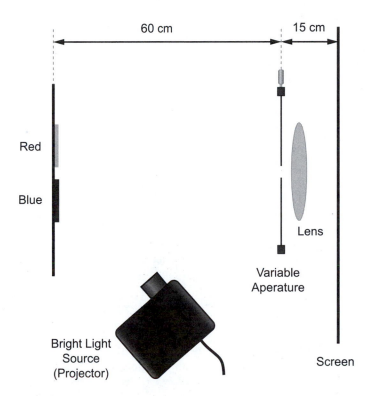

b) After making the image as clear as possible, predict how the image will change (if at all) if you further enlarge the aperture.

c) Now slowly increase the size of the aperture. Describe what happens to the size and appearance of the image on the screen as the aperture becomes larger?

_____

The ability of our model eye to cast a small, sharp, image of an object on the screen is rather impressive. In fact, if you haven't already tried it, use your hand as the object. The image is absolutely fantastic! Seeing such an accurate image on the viewing screen provides powerful evidence that the light that is scattered from the object does in fact carry information about that object. We might therefore conclude that the information carried by the light can be passed on to us if this light happens to fall upon our eyes, which organize this information into an image, presumably in a similar manner to that of our simple model. Interpretation of this image is left to the brain, which receives the information via the optic nerve. There remains one major question left to answer. How, exactly, does the lens organize light into an image?

2.2   CHANGING THE DIRECTION OF LIGHT

We began the last series of activities by creating a simple hypothetical model of the eye consisting of a hole to represent the pupil and a screen to represent the retina. This simple model created focused images of brightly illuminated objects when the hole was very small. When the object was poorly illuminated, however, a larger diameter hole was needed to cast more light on the screen, which led to a fuzzy image. When we added a glass lens to our eye to create a more sophisticated model, we saw that the image became sharp again. How does the lens accomplish this task? What is the lens doing to the scattered rays of light in order to create a sharp image of the object?

How Does a Lens Sharpen an Image?

The sketch that follows shows three of the many rays of light scattering from a tiny red spot on an object we are trying to "see" with our more sophisticated model of the eye. The dotted segments of the light rays show where each ray would have gone in the absence of the lens.

Activity 2.2.1 How Does a Lens Work?

a) On the following diagram, draw in where the three light rays shown *must* travel if they are to create a sharp image of the tiny dot on the screen.

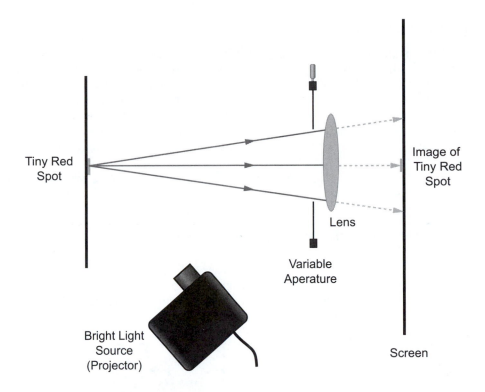

b) Do you think any of the light rays will change direction as they
    pass through the lens? If so, do they change direction in the same
    way? If not, can you give some kind of rule that might explain how
    a ray changes directions?

c) Can you think of any way that the shape of the lens might account
    for the bending behavior observed? Explain briefly.

Why Does Light Bend?

While it may seem obvious that the lens does in fact cause some light rays to bend, you probably noticed that they don't all bend in the same way. In fact, one of the rays in the above diagram doesn't bend at all. What determines the way these light rays bend? To try and understand this question, we will focus our attention on an investigation that will allow us to learn about the behavior of light as it enters or leaves a material like glass. For simplicity, we will be using a rectangular piece of Plexiglas—a material that has optical properties similar to those of glass (or water). Because we will be dealing with light rays that enter the Plexiglas at different angles, it is very convenient to use a normal line to as a reference for these angles. If you are uncomfortable with your understanding of normal lines, you should review Activity 1.3.1.

Activity 2.2.2 Light Entering Plexiglas

a) The following diagram shows a beam of light entering a piece of Plexiglas at an angle. Make a rough sketch of your prediction as to the path of the light ray as it enters, travels through, and then exits the Plexiglas. Use a ruler to draw precise lines.

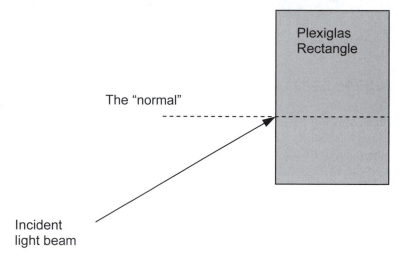

b) Now, send a single beam of light from a ray box into a piece of Plexiglas. Accurately sketch with a ruler your Plexiglas rectangle and the light ray as it enters and leaves. First, let's concentrate on the light ray as it *enters* the Plexiglas. Draw a normal line that coincides with the point where the light beam enters the Plexiglas. Does the light ray change direction so as to be closer to or further away from the normal line (as opposed to if it hadn't changed direction at all)?

c) Aim the incoming light beam so that it strikes the Plexiglas at different angles. Does it always bend in the same direction as it enters the Plexiglas? Does it always bend by the same amount? Is there any angle in which it doesn't bend at all? Explain in your own words this behavior. (You might want to use a protractor to help convince yourself what is happening.)

d) Now let's focus our attention on what happens when the light ray *leaves* the Plexiglas. Accurately sketch with a ruler your Plexiglas rectangle and the light ray as it enters and leaves. Draw in a normal line that coincides with the point where the light beam leaves the Plexiglas. Does the light ray change direction so as to be closer to or further away from the normal line (as opposed to if it hadn't changed direction at all)?

e)   Compare the behavior of light passing from air into Plexiglas with
     that of light moving from Plexiglas into air. How are the behaviors
     similar? How are they different?

Checkpoint Discussion: Before proceeding, discuss your ideas with your instructor.

Light travels inside Plexiglas as it travels in air—in a straight line. Nevertheless, as you
undoubtedly noticed, something interesting happens at the interface between the
Plexiglas and the air. The fact that light changes directions at an interface (boundary)
between two materials is called *refraction*. Thus, when light travels from air to glass,
from glass to water, or from water back to air, some refraction occurs. The actual amount
of the bending depends upon the specific materials and angles involved.

You now have the basic knowledge needed to understand how simple lenses work. The
following activity should make this clear

Activity 2.2.3 Understanding a Simple Lens

a)   Sketch the path of the light beam in the following figure. The beam
     begins in air, travels through some glass and ends up back in air.
     Draw a normal line where the beam enters the glass and show the
     how you expect the light ray to bend towards or away from the
     normal at that point. Continue the light ray's path straight through
     the glass. Draw a normal line through the point where the ray exits
     the glass and show roughly how the ray will bend as it passes back
     into air. (If you need help, try passing a light ray through your
     Plexiglas rectangle.) Use a ruler to draw precise lines.

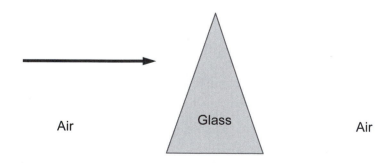

b)   Repeat your sketch for the following situation.

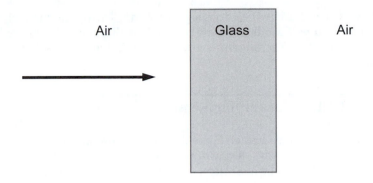

c)   And lastly, sketch the predicted refraction for the following situation.

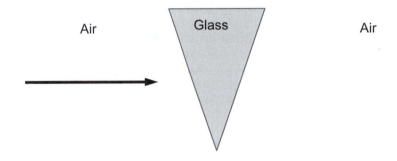

d)   Now let's put all this all together. A converging lens can be thought of as being made up from the three sections you just studied (a real converging lens would be nice and smooth, but this works as a simple model). Shown below is a sketch of this kind of a lens, along with three light rays traveling towards the lens. Using a ruler, make a rough sketch of the path of these light rays (you should use the results from the last three questions to roughly determine what the path of these light rays will be). Why do you think this kind of a lens is called a converging lens?

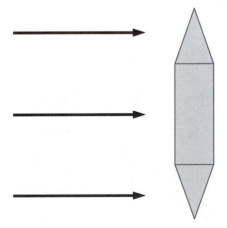

2.3   A LENS IN ACTION

Now that you have some idea how a converging lens works, lets try out the real thing. For this next activity you will use the ray box so that 5 small beams of light are being emitted parallel to each other. These beams should look similar to the rays drawn in the sketch in the previous activity.

Activity 2.3.1 A Converging Lens

a) Place the converging lens on top of the sketch below and orient the ray box so that the beams of light go through the lens approximately as shown. Sketch how each beam changes direction both when it enters and exits the lens. Do the light rays behave as you predicted in the last activity?

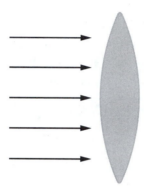

b) In most cases, the light rays from an object do not travel parallel like those from the ray box in the previous question. Shown below is a 2 to 1 scale drawing of a 3 1/2" golf pencil and a converging lens. Draw 4 or 5 of the many light rays that leave the tip of the pencil and hit the lens. How do these light rays differ from the ones produced by the ray box? Sketch your prediction for what the path of the light rays will be after they pass through the lens. (Use a ruler to draw precise lines.)

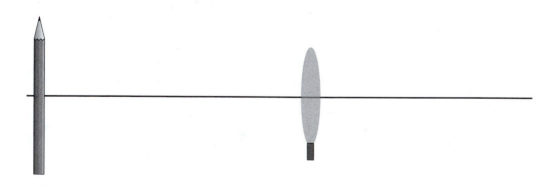

c)   Will light that scatters from the tip of the pencil hit *every* part of the lens? Will light from other parts of the pencil also hit every part of the lens? Explain.

d)   To find out what happens to light from the tip of a real pencil, perform the following experiment. ***On the following page*** is a full size sketch of a golf pencil and the two-dimensional converging lens, about 15 cm apart. Place a pencil and lens in these positions so that they don't move. Tape the pencil in place and sketch in the position of your lens so that if it moves, you can replace it. Arrange the ray box so that a single thin beam of light grazes the tip of the pencil and hits the lens (avoid the extreme edges of the lens). Sketch the path of the beam going into the lens and coming out of the lens, all the way to the edge of the paper. Next, rotate the ray box so that the beam still grazes the tip of the pencil, but hits a different spot on the lens. Sketch the path of this beam. Do this for 4 or 5 different beams of light. What happens to light from the tip of the pencil that passes through the lens? Repeat this for light that scatters from the other end of the pencil. When you are finished, use a ruler to do a careful scale drawing on the following half-scale diagram of the light beams you traced on the big piece of paper.

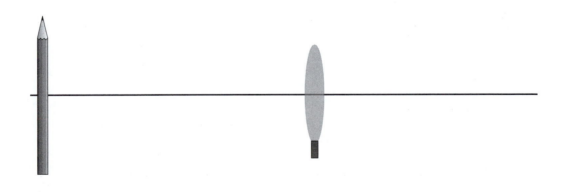

A More Careful Study of Your Model Eye

The previous activity should help you see how a lens can focus many light rays coming from one part of an object to a specific spot, while simultaneously focusing many light rays coming from a different part of the object to a different spot. In fact, light rays coming from every part of the object are being focused to different spots in a similar manner. It is quite amazing that the lens can focus light from so many different places, yet keep things organized so an image is formed!

Now it's time to come back to the question posed earlier in this section. Set up the system you used in Activity 2.1.5. That is, place an aperture, a lens, and a screen on an optical bench. Set an object about 60 cm from the lens and illuminate it with a MagLite flashlight.

Activity 2.3.2 What About a Cat's Eye?

a)   Keeping everything else at a fixed position, slowly move the object until you see a sharp image on the screen. Predict what will happen to the image if you cover up half the lens. Will it matter if you cover up the right half or the left half? **Hint:** Is light from every part of the object hitting every part of the lens?

b)   Now cover half the lens with a piece of paper and record your observations. Can you explain what you observe?

c) Explain what happens to the image if you try covering up different parts of the lens. What does this imply about the shape of the pupil? Would we see the world differently if our pupils were slits (like a cat's) instead of round? Explain briefly.

| Checkpoint Discussion: Before proceeding, discuss your ideas with your instructor. |

How the Eye Adjusts

Your model eye, consisting of an aperture, lens, and screen, works very well at creating a small, sharp image of an object at a particular distance. But as you well know, our eyes are capable of focusing on objects at many different distances. You might wonder how a real eye avoids this limitation of our simple model. To gain some insight into this, try the following experiment. With one eye closed, hold your finger about six inches in front of the open eye. Now focus on your finger and then change your focus to something in the distance and then back again. You will notice that in fact our eyes do suffer from this drawback a bit. When looking at a distant object, your finger will appear blurry, and vice-versa. You will also notice that something allowed your eye to change your focus from your finger to the distant object. This "something" is the subject of the next activity.

Activity 2.3.3 How the Eye Adjusts

a) Set up your model eye and adjust the position of the object until you have a nice clear image on the screen. Now, try to find any other locations of the object that will give you a clear image. To do this, slowly move the object toward the lens and away from the lens keeping everything else fixed. Describe your observations.

b) Place the object so that you have a sharp image. Then, get another lens from your instructor, one that is a little thicker than the one you already have (a 5-10 cm focal length works well). Without changing anything else, replace the lens in your model eye with the second lens, and slowly move the object until the image on the screen is in focus. How did you have to move the object in order to focus the image? Is there any difference in how the image looks now compared to when the original lens was in place? Explain.

c) Our eyes are not capable of changing lenses each time you want to focus on a different object On the other hand, they are capable of re-focusing. How do you think your eyes can accomplish this task? **Hint:** Refer back to Figure B-6, which shows a cut-a-way diagram of the human eye and note that the lens is not rigid like glass, but rather slightly deformable.

d) Some people's eyes are not capable of focusing on objects that are too close or too far away. Explain how using a lens in front of the eye might help these people to see better. (If you think it will be helpful, feel free to sketch a diagram.)

The Real Eye

Your physical model for the eye is admittedly a bit crude. Making a working model doesn't prove conclusively that the eye works like an optical lens. But your observations on how lenses work, combined with medical pictures of the eye, provide powerful evidence that the eye does organize light into an image on the retina. There exists a scientific principle known as Occam's Razor that instructs one to make as few assumptions as possible. If two theories equally well explain an observation, Occam's Razor advises us to side with the simpler model. This does not mean that all complicated models are wrong; just that we should not make things unnecessarily complicated. Your model of the eye is elegantly simple. Light scatters off of objects, travels to the eye, and then an image is formed on the retina. The optic fiber then conveys this image to the brain for interpretation. Should an observation be found to contradict this simple model then we might look for more complicated solutions. For now, however, there is no reason to look beyond this simple model.

A slightly more detailed side view of the eye is shown in Figure B-8. Note the large curvature of the cornea. You can also observe this by looking at a friend's eye from the side. From your earlier observations you know that when light hits a curved interface it bends. It may not surprise you then, to learn that most of the bending of the light that hits our eye occurs not from the lens, but rather from the very-curved shape of the aqueous humor. The aqueous humor is a liquid region that bends the light that hits it towards the pupil. The lens in the eye is used primarily for minor adjustments necessary to focus on objects at different distances. This is accomplished with the aid of the ciliary muscles, which subtly reshapes the lens so as to produce clear image.

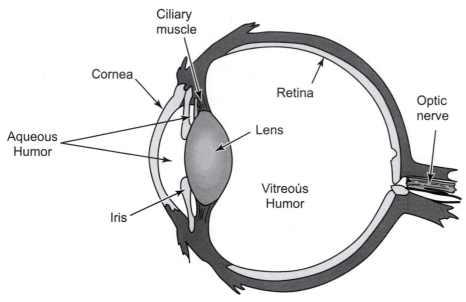

Figure B-8: A slightly more detailed cutaway sketch of the human eye showing all the main components.

Despite our model's crudity, it does capture the essential features of how our eyes work. The aqueous humor, lens, and vitreous humor act together on light in exactly the same way our single lens does in our model eye. Scientists often begin with a simplified model of the phenomena they are investigating in order to get an understanding of the central features. Then, once they understand these elementary properties, they introduce more subtle additions to the model to help account for more of the observed features. Generally, these features are less dramatic than some of the basic features and are also typically more complicated. Thus, the model gets more and more complicated as your understanding gets more and more thorough.

3	*COLOR: HOW CAN WE SEE IT?*

In the preceding sections we have investigated how information from the outside world might reach our eyes and also how our eyes might organize this information. Our model of the eye as a simple lens system, along with the idea that light scatters and can travel to our eye, does explain how information such as the shape, size, and texture of an object might be conveyed from the object to your brain. These are geometric properties, and it isn't too far fetched to believe that light, which travels in a straight line, might keep this information as it travels.

A feature that we haven't discussed, however, is color. What *is* color? Why do we see some things blue and others red? Is color a property of light, of physical objects, or of both? We are familiar with colored light and certainly we see colored objects, but is there a relation between the color of light and the color of an object? You may have noticed that our model eye reproduced colors beautifully. But this doesn't tell us much about why objects appear colored. Trying to understand color will be the subject of this section.

You will need some of the following equipment for the activities in this section:

- Set of high quality RGB and CMY filters for demo [3.1]
- Prism or Plexiglas trapezoid [3.1]
- White light ray box [3.1]
- Colored filters and gels for student use [3.1]
- Diffraction grating spectrometer [3.2]
- Flashlights [3.2]
- Crayons (Red, Orange, Yellow, Green, Blue, and Violet) [3.2]
- Optional: Gas discharge light sources (H, He, etc.) [3.2]

3.1   COLOR AND LIGHT

We will begin by investigating whether light itself has color. Can one change the color of light and, if so, how? Can you turn red light into blue light? Green light into orange light?

Activity 3.1.1 A First Look at Color

a)   Discuss amongst your group how you might turn white light red. Do you believe it is possible? What about turning red light blue? Be prepared to discuss your conclusions with the rest of the class.

b)   Your instructor will demonstrate some aspects of color by placing various combinations of filters on an overhead projector that sends out white light. For each case, predict the color of light you will observe on a screen. Record the color actually observed. Were you correct? There are extra rows in the table for additional combinations you or your instructor might want to look at.

Filter Combination	Predicted Color	Observed Color
Red		
Red + Blue		
Magenta + Cyan		
Blue + Cyan		

c) Based upon your observations, can you turn red light blue? Cite evidence supporting your position. Does this imply that the light itself is becoming colored as it passes through a filter or not? Explain briefly.

Ptolemy once said, "color is not seen unless light cooperates." In the above activity, light certainly didn't seem to cooperate in a predictable manner! The observations you made above are difficult to reconcile with our everyday experience. This is because few people have direct experience with colored light. After all, we are surrounded with white lights. Light bulbs and fluorescent lights are white, and even the sun is practically white. Yet when white light shines on certain objects we often see a color other than white. Our investigation of light and color begins with white light, and asks the question: "What is the color *white*?"

In investigating white light, scientists as far back as the 17th century made use of a triangular piece of glass known as a prism. The prism has a startling effect on white light that passes through it, one that you no doubt have observed before.

Activity 3.1.2 "White" Light and Prisms

a)   Below is a sketch of a light ray entering a prism. Use what you
     learned about the refraction of light through Plexiglas to make a
     rough sketch of the path of the light ray as it passes through the
     prism. **Hint:** Draw a normal line that passes through the point
     where the light ray hits the prism. What do you think you will
     observe when the white light exits the prism?

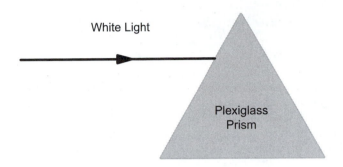

White Light

Plexiglass
Prism

b)   Place the flat side of a prism on the table and use the ray box to
     shine a single, narrow beam of white light into it at about the angle
     shown above. Slowly twist the prism back and forth until the
     exiting beam of light is spread out as much as possible as it leaves
     the prism. Sketch what you see.

c) You should have seen that the emerging light has several colors.
    What do you think the prism is doing to the incoming light? Do
    you think the color is in the light before it reaches the prism or is
    the prism adding the color to the beam? Explain.

d) Place a two-dimensional converging lens behind the prism so that
    the light passes through the lens after it passes through the prism,
    as in the figure below. Rotate the lens back and forth until the
    emerging beam becomes as narrow as possible. What happens to
    the light after it passes through the lens (at the narrowest part)?

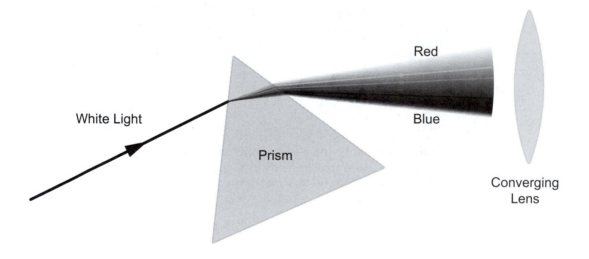

Where do Colors Come From?

When white light refracts through a prism, it spreads out into a "rainbow" of colors.
When this rainbow is refocused by a lens, the colors combine to again form white light.
This raises many questions. Is all white light composed of colored light or is there a
separate color "white?" Is the light that enters the prism colored even though it appears

white? Is the prism "coloring" the light somehow similar to the way a crayon changes the color of white paper? How do blue light, red light, and white light differ?

We began this section on color by observing the combined effect of various filters on the color of light. At that time we did not attempt to understand how the filters were affecting the light. The natural idea that the filters "color" light is only partially correct, as you discovered when you attempted to use a blue filter to turn red light blue. The following activity has a dual purpose. First, you will investigate the light that passes through the filter in an attempt to answer some of the questions raised above. Second, you will learn more about how filters affect light that passes through them.

Activity 3.1.3 Colored Light and Filters

a) Predict what you will observe when you put a red filter in the path of the beam after it exits the prism as in the following sketch. What do you think you'll see after the light has passed through the filter?

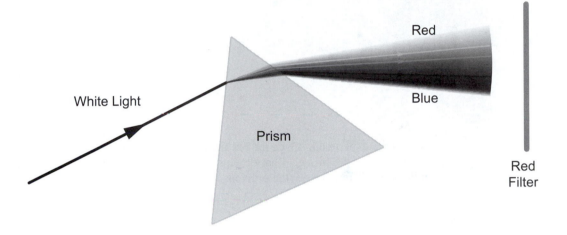

b) Now try the experiment. Place the red filter in the path of the beam after it exits the prism. What do you observe? What do you think happened to the green and blue light that left the prism?

c) How does a red filter affect a "rainbow" of light?

d) Now predict what you will observe if you place the red filter in the path of the light beam before it enters the prism, as shown in the following sketch.

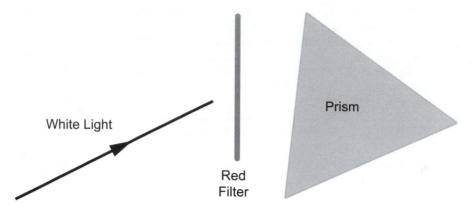

e) Now try the experiment. Place the red filter in the path of the beam before it enters the prism. What do you observe? Is the end result similar to or different from the previous experiment?

f) Do you think the white light entering the prism contains colored light? What evidence leads you to make this assertion?

**Checkpoint Discussion: Before proceeding, discuss
your ideas with your instructor.**

Are all Colors in the Spectrum?

In the last activity you observed that a prism appears to separate white light into different colors, called a *spectrum*, and that a colored filter prevents the full spectrum from appearing. You may also have noticed that some everyday colors were not present in the spectrum. Brown, for example, is conspicuously absent from the spectrum, as are gray, magenta and cyan. Nevertheless, if you look around you will certainly perceive some objects as brown or gray. Where do they come from if they are not in the spectrum?

One of the more remarkable aspects of vision is our ability to perceive a combination of light not as the sum of its parts, but as a distinctly different color. For example, when you used your lens to refocus the spectrum back down to a spot you saw it not as a jumble of red, orange, yellow, green, and blue, but as white. We will conclude our investigation of colored light by looking at exactly what makes up different colors of light.

3.2 TRANSMISSION GRAPHS AND THE SPECTRUM OF LIGHT

We will use a *diffraction grating spectrometer*[4], rather than a prism, to reveal the spectrum of light. A diffraction grating causes the colors in the spectrum to "bend," just like in a prism, but is much easier to use. Your instructor will show you how to use a spectrometer, but essentially you hold it up to your eye and look slightly to the side of a light source.

Representing Spectra of Light

When you look at a light through the spectrometer you see the colors of the spectrum. If you look at the sun or a white incandescent light source (a "normal" light bulb) through a diffraction grating, the spectrum will look like a "full rainbow" (all of the colors are bright). We can represent this spectrum with the following graph, which shows that each of the "rainbow" colors (Red, Orange, Yellow, Green, Blue, Indigo, and Violet) is present.[5]

Figure B-9: A transmission graph showing a "full rainbow" of colors.

If you pass this white light through a colored filter and look at the spectrum, you would see that some of the colors are absent, or at least not as bright as some of the others. In this case, you might make a sketch that showed only some of the colors being present. An example is shown in Figure B-10, which shows a spectrum that has light in the blue-green part of the spectrum but nothing else. This kind of graph is called a *transmission graph* because it shows which portion of the spectrum has been transmitted through the filter.

Figure B-10: A transmission graph showing only the blue-green portion of the spectrum.

---

[4] An optical *spectrometer* is any device that separates the colors of light so that you can inspect the spectrum (color distribution) of the light. Both the prism and the diffraction grating are used as spectrometers in this unit.

[5] Of course, if you look carefully, you will see that these colors blend smoothly from one to the other. There are, in fact *many* more colors present than the seven listed here. We list he colors of the rainbow as ROYGBIV for convenience.

Finally, sometimes the colors do not form broad bands but instead they form rather sharp lines. In this case, we draw a thin line indicating the color. A spectrum with three narrow lines, one orange, one yellowish-green, and one slightly darker than indigo, is shown in Figure B-11 below.

Figure B-11: A transmission graph showing three sharp lines.

Activity 3.2.1 The Primary Colors

a) Look through your spectrometer at "normal" incandescent light, such as the flashlight you have been using. Using either crayons or colored pencils, fill in the transmission graph below.

"Normal" Light

b) Make a rough sketch of the transmission graph when using your red, blue, and green filters in front of your flashlight.

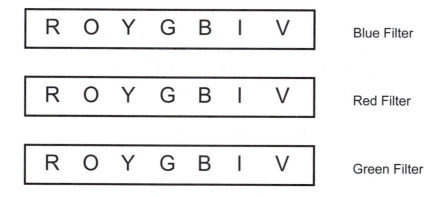

Blue Filter

Red Filter

Green Filter

Primary Colors

You may have noticed that the transmission graph for the red filter has a higher intensity in the red-orange portion of the spectrum (the bottom third, roughly), and lower intensities everywhere else. The green filter has a higher intensity in the yellow-green portion of the spectrum (the middle third, roughly), and lower intensities everywhere else. The blue filter has a higher intensity for the blue-indigo portion of the spectrum (the upper third), and lower intensities everywhere else. Because of this, scientists often speak of the red, green, or blue portions of the spectrum. In fact, one can define *primary* colors (red, green, and blue) as those colors that have *only* the bottom, middle or upper third of the spectrum in them. The plastic filters you have been using transmit more than just the primary colors. Your instructor may have a set of precision-made filters to demonstrate what these "colors" really look like.

Notice that the three primary colors each contain about ⅓ of the full spectrum and that they don't overlap with each other. It seems reasonable then, that if one could somehow arrange to "add" these three primary colors together, that you would end up with the full spectrum, or white light. In fact, this kind of color addition can be used to create millions of different colors and is used in television and computer screens. Note: If you use a magnifying glass to look very closely at your television or computer screen you'll see lots of small red, green and blue dots. We will explore this topic later on in this unit.

As another example of spectra and transmission graphs, your instructor may have a sampling of gas light sources such as hydrogen, helium, neon, etc. These light sources have very interesting spectra. If so, complete the following activity.

Activity 3.2.2 Spectral Fingerprints

a) Examine two light sources and describe their color. Then, look at the spectra and make a rough sketch of it below. Don't forget to indicate what your light source is.

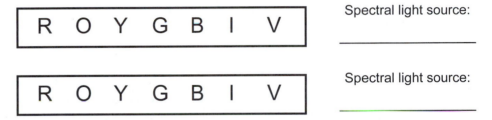

Spectral light source:

_____

Spectral light source:

_____

b) Describe how these spectra differ from the "full rainbow" you observed earlier.

You should have noticed that the spectra of these light sources are quite different from each other. In fact, every gas can be excited to emit light like this when a high voltage is applied, and each gas has its own unique *spectral fingerprint*. Astronomers use these fingerprints to identify the materials that are present in stars by analyzing their spectra. For example, the lines observed in the spectra of a star are compared to the spectra of specific elements we've observed here on Earth. These lines and their relative intensities give astronomers information about which elements these distant stars are made of.

**Checkpoint Discussion: Before proceeding, discuss
your ideas with your instructor.**

3.3   REAL OBJECTS

So far we've been dealing with light, be it colored or white (by now you should realize
that such a distinction is essentially meaningless). Now we will try to answer our original
question of why we see color in objects. Is the color a property of the object, of the light,
or both?

Activity 3.3.1 What Happens to the Colors?

a) When you look through a green filter at a light source, the light
    looks green. Since the white light actually contains a "rainbow" of
    colors, explain which colors make it to your eyes and what
    happens to those that don't.

b) Now imagine looking at a white object in a room filled with white
    light. Describe the light that reaches your eyes. What if you were
    to look at this object through a green filter? What light would reach
    your eyes?

c) Now imagine looking at a green object in a room filled with white
    light (without looking through any filters). What do you think is
    the color of the light that reaches your eye? Explain.

d) Since white light is made up of the full spectrum of colors, what happens to the non-green light that hits this green object? Does any of it reach your eyes? Explain.

e) What color do you think a primary green object would appear to be if you were to view it through a primary red filter? Explain.

f) What color do you think this primary green object would appear to be if it was illuminated with only primary blue light? Is this at all similar to looking through a blue and green filter at the same time? Explain.

_____

If you had any trouble with the above activity, perhaps it will help to think in terms of the light that reaches your eyes. If something appears green, or blue, or red, or some other color, what does that tell you about the light that is reaching your eyes? Of course, the light that is shining on the object also plays some role, for you have already observed that red light does not contain the whole "rainbow." These next few activities deal with the question of colored light on colored objects. **Note:** Even though there are many different shades of red, green, and blue, when we refer to these colors throughout the rest of this unit, we mean primary red, primary green, and primary blue.

Activity 3.3.2 Colored Light, Colored Objects I

a) A red object and a white object are set up in a completely darkened room. Recall that you cannot see anything in a completely darkened room. Now, imagine shining a red light on these objects. What color will the objects appear? Sketch the path of several light rays scattering off the objects, and indicate the color of this scattered light.

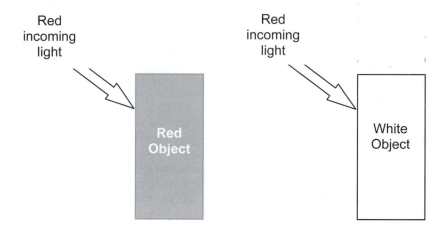

b) Based on the above scenario, do you think your eye would see a difference between these two objects? Explain why or why not

c) Now, let's try the experiment. Place a sheet of red cardboard partially over a similar piece of white cardboard in a darkened room. Shine a narrow-beamed flashlight through a red filter so that the bright spot is on both pieces of cardboard at once. Comment on what you observe compared to what it looks like when the flashlight is not going through a red filter. **Note:** If possible, the red filter and the red cardboard should be the same shade of red. This experiment should be done in a *very* dark room to reduce contamination by outside (not red) light sources.

d) Based on these observations, explain how someone might mistake a white object for one of a different color. This could cause some confusion when discussing the color of an object with a friend. How could you *define* the color of an object so to avoid this confusion? Write your definition below.

If the color of the filter and the color of the paper used in the previous activity weren't exactly the same, you may have noticed a slight difference between the red and white objects when illuminated with red light. The more closely the color of the object and the filter match each other, the more difficult it is to tell the red object from the white object. Unfortunately, most colored filters and colored pieces of paper have a reasonably large spectrum of colors in them. Thus, even though an object looks red, it may have some blue and green and yellow in it as well. Clearly, most colors are *not* primary colors. Keep this in mind when performing the following activity.

Activity 3.3.3 White Light, Colored Objects

a) White light containing a full spectrum of colors falls upon a colored (say, primary red) object and is scattered. Since it is this scattered light that makes its way to your eyes, and you see a red object, this means that red light must be coming to your eyes. What do you think happens to the blue and green light contained in the incoming white light?

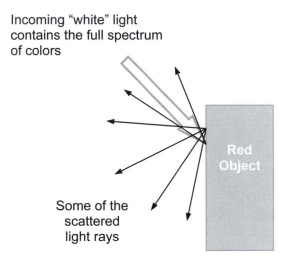

Incoming "white" light contains the full spectrum of colors

Some of the scattered light rays

Red Object

b) What do you think you would observe if you passed the white light through a primary blue filter, as shown below? Draw in some light rays and explain your answer.

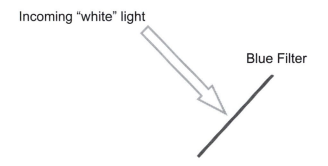

Incoming "white" light

Blue Filter

Red Object

c) Now try the experiment. You might want to try red light on a blue object, or blue light on a red object. These combinations typically work the best (can you explain why?). You can also just look through a red filter at a blue object. Play around a little, and give a summary of what you observe.

This aspect of colored light on colored objects is often surprising and confusing. As already mentioned, one reason for some of the difficulties is that most of the colors that we observe in the everyday world are not "pure." This means that when an object looks "red," there are typically more colors in it than just red (including some blue and green). Your instructor may have a demonstration of "red light on a blue object" using some precision filters. The results are much more dramatic than using the inexpensive filters.

> ## Checkpoint Discussion: Before proceeding, discuss your ideas with your instructor.

Color Subtraction

So far, we have been talking about colored filters and colored objects. You may have noticed that in both of these cases, only a portion of the spectrum that is initially present is scattered from the object (or transmitted through the filter). Thus, when looking at white light, we begin with a full spectrum of colors and end up with only a portion of the spectrum. We call this a *subtractive process*, because we only end up with a portion of what we started with. In essence, the object or filter has "removed" some of the colors from the spectrum.

When we "see" an object, what we see is light that comes from the object. This light has been affected by the object in some way. Some of the colors are scattered, and some are absorbed. Thus, the light coming from an object carries *information* about the object (i.e., which colors it scatters and which it absorbed). Our eyes then organize this light through the lens/pupil/retina combination to form an image of the world so that our brains can make sense of it. But as mentioned in the introduction, we will not pursue the question of how our brains make sense of this information.

Color Addition

We will complete this section with an activity that brings our discussion of color full circle. Since colored objects and filters subtract out some of the colors, you might be wondering if colors can also be added together. This is the topic of the following activity. **Note:** This activity may be done as a demonstration by your instructor.

Activity 3.3.4 Adding Colors

a) Consider the following situation where we shine three narrow
    beams (they do not overlap) of different colored light onto a white
    object in a completely darkened room. If you were looking at this
    object, describe which colors are being scattered and what you
    would observe. Does it depend on where you look? Explain
    briefly. **Hint:** What portion of the spectrum does a white object
    scatter?

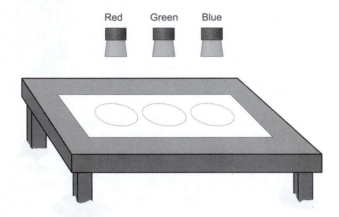

b) Now imagine that all three colored lights are shining on the same
    spot, as shown below. Describe the scattered light and what you
    think you would observe. Does it depend on where you look?
    Explain briefly.

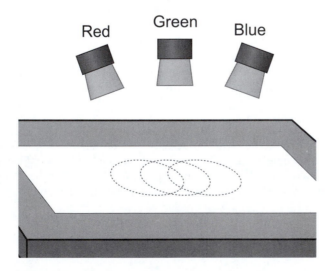

c) What do you think you would see if only two of the three lights were shining on the same spot? Explain

d) Now try the experiment. Remember, you need to shine different colored light onto a white object in a *completely darkened room*. Describe what you observe. Were you surprised?

If you are not able to carry out these observations due to equipment limitations, your instructor should be able to demonstrate the various situations for you. This is one of the more interesting and "colorful" things you can do with color.

4	*THE REAL WORLD—RAINBOWS, BLUE SKIES, AND SUNSETS*

Now that we know quite a bit about light and color, we will consider a few of the most common natural occurrences that involve light and color. What makes a rainbow? Why is the sky blue? Why does the sky turn reddish-orange when the sun sets? These are phenomena that are so common, most people just take them for granted without giving them much thought. In this section, we will consider these phenomena from a scientific perspective using the knowledge we have gained so far.

You will need some of the following equipment for the activities in this section:

- White light ray box [4.1]
- Petri dish [4.1]
- Tank of water with powdered creamer [4.2]
- MBL system [4.2]
- Colorimeter [4.2]

4.1   RAINBOWS

We begin with perhaps the most beautiful, naturally occurring phenomena involving light and color—the formation of a rainbow. After having learned some of the basic features of light and color, you are now in a position to analyze how rainbows are formed from a scientific perspective.

Activity 4.1.1 Initial Thoughts

a)   Do you think it is possible to see a rainbow while standing in the rain? Do you think it is possible to see a rainbow when you are not standing in the rain?

b)   The word "rainbow" suggests that rain is needed to form a rainbow, but sunlight is also necessary. As we've already seen, light is necessary if we want to see anything. For a rainbow, the source of this light is usually the sun. Below, make a sketch indicating where you think you, the sun, and rain must be in order for you to see a rainbow. Also draw the path of the light from the sun to your eyes.

As always, we would like to begin with the simplest model we can think of. This allows us to focus our attention on the most important features of the phenomenon. If needed, we can always add more complexity to our model. To begin, we must ask ourselves what is absolutely essential in the formation of a rainbow. Certainly, the sun and some rain are necessary. There may be other factors that influence the rainbow's brightness and size, but let's see what we can learn by considering only the sun and the rain.

Since we are assuming both the sun and the rain are essential ingredients, it seems pretty obvious that the sunlight must interact with the raindrops in some way. So we will begin by shining light from our ray box (to model the light rays from the sun) into a circular Petri dish filled with water (to model a raindrop). Whenever we try and model a physical phenomenon, there is always the possibility that the results of our model won't quite agree with the physical situation. As previously mentioned, we are only interested in understanding the primary ingredients in the formation of a rainbow. If there are small differences between what we see using our model and what is observed in reality, we can address those issues later.

Activity 4.1.2 Making a Rainbow

a) Begin by filling the Petri dish about half full of water and placing it on top of a piece of white paper. In a darkened room, use your ray box to send a wide beam of light into the side of a Petri dish filled with water. Then, *slowly* move your slice of "raindrop" up and down through the light and look carefully on the white paper for any exiting light that resembles a rainbow. **Note:** The room lights should be dimmed for this experiment. You will need to look carefully; since we are only dealing with one raindrop, the rainbow will be small.

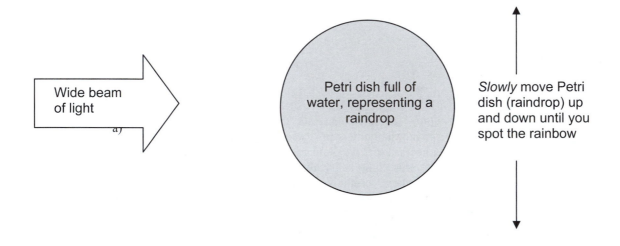

b) Once you have found the rainbow, try to determnine the path that the light is taking, inside the raindrop. You can do this by sending a thin beam of light into the drop and moving it until you reproduce the rainbow. (If you're having trouble, try sending the beam in at a grazing angle.) Then, look at the direction of the incoming light ray and look at the direction of the outgoing light ray and try to deduce the path in the raindrop. (You might be able to see the path of the light beam on the white paper. If not, you can put your finger in the water and follow the path by keeping the light on your finger.) When you understand the path of the light, make a reasonably accurate sketch below (use a ruler) showing the complete path of the light ray.

c) Clearly, we can't see a rainbow from everywhere, otherwise we'd see a rainbow whenever it rained. So, where do we have to be and where do we have to look in order to see a rainbow? Shown below is a sketch of some raindrops and incoming light rays (from the sun) with a person in five different locations. Based on your results from part b), indicate in which of the five positions (A-E), the person will be able to see a rainbow. For those who can see the rainbow, where in the sky should they look? Should they be looking at the same part of the sky? Explain why or why not.

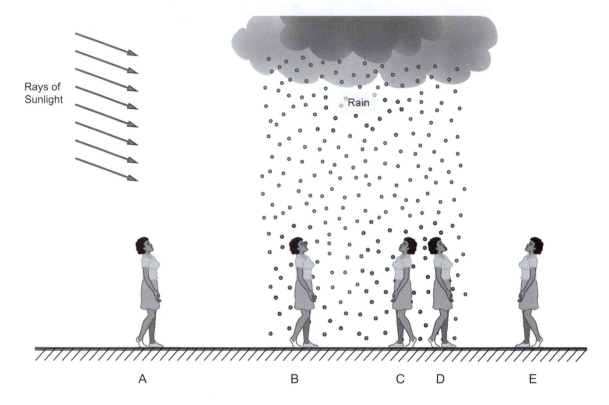

Now that you understand how a rainbow forms, the next time you see one, you should take a little time to contemplate everything you have learned about light and color. Hopefully, this will add to your enjoyment of the rainbow.

4.2   THE SUN AND THE SKY

"Why is the sky blue?" is probably one of the most frequently asked questions of all time. Although this is a fairly complicated phenomenon, your understanding of light and color is solid enough to attempt an explanation to this age-old question.

Activity 4.2.1 The Color of Air

a)   Can you see air? If so, what color is it?

b)   Can you see the sky? What color is the sky?

c)   If you can see the sky, and the sky is nothing more than air, does that mean you are seeing the air? Why or why not?

d) What about at night? What color is the sky at night? Why do you think the color of the sky changes between day and night? What clues does this provide you regarding the blue sky? Explain briefly.

In the beginning of this unit, you saw a demonstration of how a light beam (flashlight or LASER light) was invisible in air unless there was something to scatter it to your eyes (such as powdered milk in water, or chalk dust in air). The next activity uses this concept to try to explain the color of the sky.

Activity 4.2.2 Atmospheric Scattering

a) On a clear sunny day when the sun is high in the sky, if you look away from the sun, you will see a nice blue sky. This is a clear sign that blue light is coming to your eyes. What is the *source* of this light? **Hint:** What color is the sky at night?

b) If you were foolish enough to look directly at the sun when it is high in the sky (which you should never do), you would see that it appears slightly yellowish. Assuming that the light emitted from the sun is perfectly white, what color has been *removed* before it reaches your eyes? **Hint:** Which of the primary colors combine to give yellow?

c) With your group, try to determine a hypothesis that can explain your answers to parts a) and b). **Hint:** Consider the possibility that the air scatters light in some way. What color and in what direction(s) should light be scattered to be consistent with parts a) and b)? Make a sketch on the diagram below showing how this scattering can account for both a blue sky and a yellowish sun.

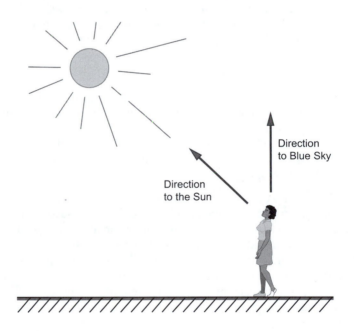

Direction to Blue Sky

Direction to the Sun

d)   You may have also noticed that when the sun is very low in the sky (sunrise or sunset), it appears orange or even red (instead of the more yellowish appearance it has when it is high in the sky). What colors have been removed from the white light so that the sun appears red? Using the ideas developed in this activity, can you account for this observation? Explain briefly. **D**do you think the amount of sky the sunlight travels through has any effect?

Your instructor will have a demonstration using milky water that exhibits the blue-sky/red-sunset scattering phenomenon. While this demonstration does not precisely mimic the conditions of the sun and the sky, it acts as a model of the atmosphere that illustrates how it is possible to scatter different colors of light by different amounts. Recall that one of the first activities you preformed in this unit was to shine a flashlight through some water with some powdered creamer in it. You may have noticed that the light passing through the milky water had an orange tint. Can you now explain why?

Measuring Color Transmission

You might be wondering if there is any way of measuring this phenomenon. It turns out that there is. The amount of blue and red light that "survives" the journey through the atmosphere can be measured with a device called a *colorimeter*. A colorimeter is a device that measures the amount of transmission of a particular color of light. In our case, we want to look at how much blue and red light survive a journey through a small sample of our model atmosphere (the milky water).

The way the colorimeter works is fairly straightforward. You have a choice of three very specific colors of light, which are aimed towards a sensor that measures light intensity. If you place something between the light and the sensor that absorbs or scatters some of the light, then only a portion of the light will get to the sensor. To see exactly how much light gets through to the sensor, we must make two measurements. The first is a measurement using clear water so that we know how much blue or red light gets through the water and the container. The second measurement will use the milky water and will tell us exactly how much of the blue or red light gets through *as a percentage* of how much got through the clear water.

For the following activity, you will use milky water from the demonstration (or you can make your own). Fill a small container with water and shine a flashlight though the water and onto a white piece of paper. Then slowly add a small amount of powdered creamer and stir the water until it is well mixed. Keep adding creamer, a bit at a time, until the light passing through the water has an orange tint to it. **Note:** Be careful not to add too much creamer. It doesn't take much.

Activity 4.2.3 Measuring Color Transmission

a) Perform a red light transmission measurement. **Note:** You will
    need to perform a measurement on clear water first, and then do
    the measurement with milky water. Print out a copy of your graph
    and write your result below, explaining precisely what it means.

b) Now perform a blue light transmission measurement. **Note:** you
    will need to perform a new measurement with clear water to
    calibrate for the blue light. Then make a measurement of the milky
    water. Write your result below. Was more blue or red light
    transmitted?

c) The light that is no longer reaching the sensor in the colorimeter is
    being scattered away by the milk particles. Using your results
    above, is there more red light or blue light in this scattered light?

d) Do you think green light is also being scattered? If so, would it be
    scattered more or less than blue light? More or less than red light?
    Explain. (If you have time, you should make a green light
    transmission measurement and check your prediction!)

e)   Discuss these results with your group and describe how they support the idea of scattering in the atmosphere being responsible for blue skies and red sunsets.

One comment needs to be made. Human perception of color is extremely complicated and not fully understood. For example, in music, when two tones are played simultaneously, we hear a harmony.[6] Someone who is well-versed in music will be able to distinguish precisely what the two notes are. When viewing color, however, this is not the case. As we have seen, when viewing two colors "on top of each other," we see a completely different color for which it is not in general possible to distinguish the constituent colors. To further complicate matters, our eye's sensitivity actually depends on the color of the incoming light, something we have not considered at all. This does not mean that our conclusions are invalid, but it does mean that we shouldn't push them too far. There are a number of optical phenomena that cannot be explained with the simple ideas we have developed. This is what makes the study of light (and the sciences in general) so exciting!

At this point, we would like to share a few things about light and color that scientists have learned that we were not able to cover in this unit. The most important of these is that light is an electromagnetic wave, identical in form to radio waves, x-rays, and microwaves. The only differences between these waves are their wavelengths, and the amount of energy they contain. The wavelength of visible light is about 400-700 nanometers (a nanometer is one billionth of a meter). You might have noticed that there are numbers inside your diffraction grating spectrometer. The spectrometers are calibrated so that these values correspond to the wavelength of the light below the number.

The atmosphere blocks a large portion of the light spectrum (which consists of all electromagnetic waves). This is important because there is a large amount of harmful radiation (electromagnetic waves are also referred to as electromagnetic radiation) that is blocked by the atmosphere. Nevertheless, a small "window" in the atmosphere allows visible light (as well as some other wavelengths) to penetrate. Presumably, our eyes have evolved to be sensitive to this portion of the electromagnetic spectrum because of this window in the atmosphere!

This finishes our classroom study on light and color. Although we have studied only a small number of phenomena that deal with light and color, we hope that your understanding of this aspect of the physical world has deepened. Also, we hope you have noticed an increase in your confidence about performing scientific work and the process involved in inquiry based learning.

---

[6] In fact, this is not always true. Complex tones are made up of multiple frequencies but sound as if they are just one frequency. This phenomenon is similar to our perception of colors and is why we are able to distinguish people's voices so easily without seeing them.

5	*PROJECT IDEAS*

It is now time for you to take on the role of scientific investigator and to design a research project focused on some aspect of this unit that you found particularly interesting. On the pages that follow, you will find a number of project suggestions. Please do not feel limited by these suggestions. You may modify any of these or come up with a completely new one on your own. We have found that many of the best projects are those dreamt up by students. We therefore encourage you to develop your own project on a topic that you find interesting. You should of course consult with your instructor as some projects require too much time or impossibly large resources. Nevertheless, anything involving light, vision or color is fair game. So use your imagination and have fun!

Your instructor may ask you to write a brief proposal that outlines the goals of your project and how you plan to accomplish them. You may find it helpful to refer to the project proposal guidelines included in Appendix B. Try to plan your project in stages, so that if you run into difficulties early on you will at least be able to complete the data collection, analysis, and interpretation. To this end, it is important to note that the project proposals listed here are intended to foster your creativity, not to tell you exactly what to do. In most cases, answering all the questions in one of these proposals would take far more time than you have. So, choose a few questions that interest you or generate some of your own, but try to keep your project focused.

You will probably want to keep a lab notebook to document your project as it unfolds. Also keep in mind that you may be presenting your project to your classmates, so be prepared to discuss your results, how you measured them, and what conclusions you can draw from them. You may find it helpful to look over the oral presentations guidelines and project summary guidelines in Appendix B as you work. These guidelines may give you a better idea of what is expected from a typical student project. Be sure to consult with your instructor about their requirements for your project as they may differ from the guidelines laid out in Appendix B.

Good luck, and have fun!

5.1   FOCAL LENGTH OF A LENS

Courtesy of Dickinson College Photo Archives

In today's society, there is a wide range of uses for lenses. Glasses, binoculars, telescopes, and cameras are a small sampling of everyday items that depend on a solid understanding of lens behavior. In order to build a product that meets certain specifications, a company must have more than a basic idea of how lenses work. In addition to understanding their qualitative features, they must also understand the *quantitative* aspects of how lenses operate.

Imagine you work for a company that is interested in building a device that uses lenses. Your task is to determine as precisely as you can how lenses work. You already know that converging lenses focus the light into images. But exactly where these images are formed and how this relates to the position of the object is not fully understood. Some of the questions you might want to investigate are:

1. What are some of the differences between converging lenses? How can you distinguish between them? Is there a number you can define to label them? If so, how do you determine this number and what does it represent?

2. Qualitatively, how are the object distance and image distance related? Make sure you carefully define what you mean by object distance and image distance. How do different lenses affect this relationship? Can you find a quantitative relationship between these variables?

3. How does the intensity of the image depend on its size? What do you *expect* would be the relationship between these quantities? How can you find out if this relationship holds?

4. Are there other kinds of lenses besides converging lenses? If so, how do they behave? Can you quantify their behavior in some way?

5.2   POLARIZATION

Figure B-12: The two pictures above were taken of the same scene. The picture on the left was taken with a polarizing filter, while the picture on the right was taken without a filter. What does a polarizing filter do and why does it reduce the glare from water? Why do you think fishermen often wear polarizing sunglasses?

Imagine you are working for a sunglass manufacturing company and your competitor has just released a new product called *polarizing sunglasses*. They are hyping these new glasses as the "greatest thing since sliced bread," particularly good for using when the sun is low in the sky or when spending time at the beach. Your company finds that sales are declining after the introduction of these new glasses and the president calls an important meeting.

At the meeting, the president throws down a pair of these polarizing sunglasses and frantically exclaims, "These things are going to put us out of business unless we can figure out how they work and make some of our own!" Your team has been put in charge of finding out what they do and explaining it to the rest of the company when you are done. Some of the suggestions from your co-workers at the meeting were:

1. Try to determine under what general conditions these glasses work well and also when they don't work so well. Do they really work well when the sun is low in the sky? What exactly does it mean to *work well*?

2. Come up with some ideas about how these sunglasses might work, and test your ideas in the lab and outside. If the sunglasses work well when the sun is low, what happens when you tilt your head from side to side?

3. Try taking two pieces of the polarizing material and looking at light as it comes through both of them. What happens when you twist one of them? What happens when you twist both of them?

4. Since these sunglasses are supposed to work well for fisherman, how is light reflected from water affected by these sunglasses? Does the angle that the light hits the water matter?

5.3   ULTRAVIOLET (UV) LIGHT AND SKIN PROTECTION

©Peter Cade/Stone

The presence of UV rays in sunlight is of major concern to people afraid of getting skin cancer. The loss of ozone in the Earth's upper atmosphere allows increasing amounts of this light component to reach the surface of the Earth. It has been well established that UV light causes changes in the skin that can eventually lead to skin cancer. In certain places, the problem is particularly bad due to holes in the protective ozone layer of the atmosphere. In Australia, for example, children at taught at an early age to "slip, slap and slop"—*slip* on a long sleeve shirt, *slap* on a hat, and *slop* on protective skin cream.

Imagine that you work for a consumer's watch group that is interested in evaluating the claims of skin lotion manufacturers. A major chain store, *Floor-Mart*, has just released a new skin protective lotion that they claim is "twice the protection for twice as long for only half the price" compared to typical competing products. You are asked to make preliminary evaluations about the claims made regarding Floor-Mart's new skin product. In order to carry this out, you need to determine how much UV passes through a coating of various protective products. You may also want to evaluate the "staying" power of the creams when exposed to water, dust, etc. Your plan of action is as follows:

1.   Decide on how you might measure the amount of UV transmitted through a layer of lotion.

2.   Once you have a working detection system, measure the relative amounts of UV transmitted through a thin coating of different products. (Is it important to have the coatings equally thick? If so, how can you make sure that you have equally thick coatings?)

3.   Investigate how the product's blocking ability ages. How does water or dust affect it?

4.   Plot a graph showing stopping power versus cost and show which products give you the most protection for the buck.

5.   How does the SPF rating come into play. Does a higher SPF rating indicate that the sunscreen actually blocks more light?

5.4   PINHOLE CAMERA

©David Lees/Corbis Images

Possibly the world's simplest optical device is the so-called *pinhole camera*. One of the most frequent uses of such a device is to aid in viewing a solar eclipse. The idea is simple enough, you just make a small hole in a thin piece of cardboard or tin foil and then place a piece of frosted glass behind it (wax paper should work). The object must be fairly bright in order to see a good image (that's why it works pretty well for the sun). To help in viewing the image, you should screen your eyes from any extraneous light by covering your head with a piece of black cloth (like an old-fashioned camera).

Your objective is to determine what is going on. That is, how do these pinhole cameras work? You might try:

1. Building a few of these pinhole cameras, altering some of the variables involved. For example, try changing the size of the pinhole or changing the distance between the pinhole and the viewing screen. How do these changes affect the image?

2. Determine what makes the image bigger, smaller, brighter, dimmer, sharper, or less focused. Make some appropriate graphs to help you determine how this thing works. For example, you might try graphing the image size as a function of the distance between the pinhole and the viewing screen.

3. Once you have some quantitative data that gives you some clues as to what is happening in the formation of the image, come up with a theory as to how the camera works.

4. Try taking some actual pictures with your camera using Polaroid film. Does it work?

5.5   CURVED MIRRORS

Figure B-13: The mirror inside the Hubble Space Telescope is shaped so that light entering along the axis of the telescope will be reflected to a focus at a single point. Because the mirror on Hubble is so large, it is possible to see very dim objects. (Courtesy NASA)

The images formed by a flat mirror are common. Less common are the images formed by curved mirrors. For example, if you take a spoon and look at your reflection on the inside or the outside of the spoon, you may not know what to expect. Imagine that you were doing this and you noticed that there were some potentially lucrative uses for curved mirrors. In fact, you would like to go into business and make these mirrors to sell but you don't have the money to get started. A friend of yours puts you in contact with a venture capitalist firm which supplies money to start-up companies that they think are worthwhile endeavors. The problem is, you must convince them that you have a useful product and that you have an exceptional understanding of what is going on.

In order to get a better understanding of these mirrors, you decide to gather a few of your friends together and try and figure out exactly how they work. In order to satisfy the venture capitalists, you might want to consider the following:

1.   What can these curved mirrors do? What do you see when you look at a concave mirror or a convex mirror?

2.   Exactly what kind of uses these mirrors have and who might want to buy them.

3.   How do these mirrors work? Can you make a diagram of the beams of light from an object and show how the mirror deflects these beams to form an image? Are there any differences between the types of images seen in these mirrors?

5.6   CORRECTING VISION

©Corbis Digital Stock

Many people wear either glasses or contact lenses to correct their visions. If you look at a pair of glasses or contact lenses, however, you'll notice that they don't quite look like the converging lenses we used in class. So how do these lenses help us see?

Clearly, corrective lenses are designed to work *with* our eyes. Therefore, it may be beneficial to begin with a model of the eye and then to introduce lenses in front of the eye. You can investigate the properties of corrective lenses by themselves to see if this helps you understand how they will help fix a particular vision problem. As you work, you may want to keep in the mind the following questions.

1. What happens to the image formed by a model eye when a lens is placed in front of it? Does it appear in the same place? Does it depend on the kind of lens? Does it matter where this new lens is positioned in front of the eye?

2. How are myopia (near-sightedness) and hyperopia (far-sightedness) treated with lenses? Are the same kinds of lenses used?

3. Can the corrective lenses form images all by themselves? Try using both glasses and contact lenses.

4. What are bi-focals (and tri-focals)? Why would someone want to wear such glasses?

5.7   REFLECTION AND REFRACTION

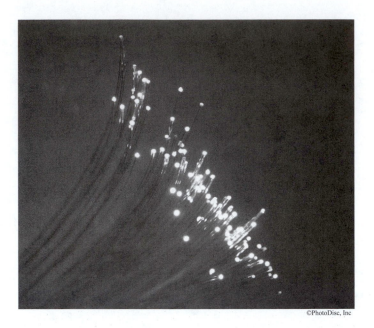

©PhotoDisc, Inc

In class, you observed that in the formation of a rainbow, light is refracted as it enters the drop of water, then it's reflected off the inner back surface, and then it's refracted once again as it leaves the raindrop. One thing that you may have noticed is that there was actually both reflection *and* refraction taking place *each* time the light encountered an interface (the boundary between air and water). Now, the refracted part of the light is transmitted past the interface while the reflected part is not. These ideas can be used to develop "light guides" (fiber optic cables) that take light from one place to another without significant attenuation.

Imagine that you were the first person to realize that these light guides had the potential to carry more information that regular phone cables. In fact, you became convinced that you could start a profitable business if you just understood how transmission and reflection at an interface depended on factors such as angle. To explore this idea, you might want to consider the following questions:

1. How does the angle of the refracted light beam depend on the angle of the incident light beam? Is this relationship the same for all materials?

2. Does the amount of reflected/transmitted light depend on the angle at which it strikes the interface? If so, how?

3. How can a beam of light be kept inside a fiber optic cable?

UNIT C

HEAT, TEMPERATURE, AND CLOUD FORMATION

DETAILED CONTENTS

UNIT C

HEAT, TEMPERATURE, AND CLOUD FORMATION

Nature, it seems, is the popular name
For milliards and milliards and milliards
Of particles playing their infinite game
Of billiards and billiards and billiards.

--Piet Hein

0	**OBJECTIVES**

1. To observe and describe temperature differences and compare these with our sensations of hot and cold.

2. To understand why a thermometer behaves the way it does and to understand the relationship between different temperature scales by constructing our own thermometer.

3. To develop a model for how thermal energy is transferred between objects and to understand how the transfer of thermal energy causes the temperatures of objects to change.

4. To build an understanding of phase changes of matter, such as the melting of ice and the boiling of water.

5. To recognize the difference between wet-bulb and dry-bulb temperature readings and understand how this difference provides a measure of humidity.

6. To use the concepts of evaporation, condensation, and relative humidity to understand the principles of cloud formation and investigate the conditions under which clouds will form.

7. To learn more about the nature of heat, temperature, and the process of scientific research by undertaking an independent investigation.

0.1   OVERVIEW

Everyday we engage in activities that are influenced by heat and temperature from drinking a glass of iced tea to feeling chilled after swimming on a hot day to stepping through a fogged bathroom after a hot shower. Even so, we rarely think carefully about the concepts of heat and temperature and how they affect our daily lives. If you want to cool down a glass of tea, how much ice do you need to add? If you add two ice cubes, will it cool the tea down twice as much as if you added only one ice cube? How much will drinking the iced tea actually cool your body. Why do you get chilled after swimming even on a hot day? What makes the temperature you "feel" seem different even if the actual temperature hasn't changed? Why does fog form in your bathroom when you take a hot shower? Will opening the window in your bathroom prevent the fog from forming or make it worse? In this unit, we will investigate the concepts of heat, temperature and humidity with the aim of learning how to answer these kinds of questions.

In addition, you will learn how these concepts fit together to explain other everyday phenomena. In the process, you will have an opportunity to work on experiments to help you understand the nature of these concepts. In the end, you will even make your own cloud. Some of your work will be done independently and some in small teams. You will likely learn the most when you are engaged in discussions with your partners. Debating your ideas will lead all of you to a more solid understanding of the concepts under study. So don't blindly work through the experiments. If you don't understand something, speak up and challenge your partners to explain it to you. The ensuing discussion will benefit you all.

The path that we will be taking in this unit is a bit complicated so it is useful to break it down into smaller, more manageable pieces. First we will look at temperature and how we measure it. Then we will investigate how two objects of different temperatures interact. Next we will study how a substance, in our case water, changes form as it is heated. This kind of transformation is known as a *phase change*. Finally, we will tie all of this together to explain one of the most ephemeral of phenomena, the cloud. This is all summarized in Figure C-1.

©Simon Fraser /Photo Researchers   ©Corbis Images

Sections 1 & 2: How does a thermometer work?

Sections 2 & 3: How are "heat" and temperature related?

Sections 3 & 4: How are evaporation and humidity related?

Section 4: What makes a cloud form?

Figure C-1: The main questions we will be tackling in this unit.

As we carry out our investigations, it helps to have some specific questions we would like answered. As already mentioned, we all know when something is "hot," but this doesn't describe things specifically enough to be useful. A natural first question might be *how does a thermometer work and what does it tell us?* As you know, you can increase the temperature of an object by "heating it up" (in an oven, for example). But how that happens may be a bit of a mystery. Therefore, another question is *how does heating something up produce a temperature change?* A third common experience for most people is the boiling of water. Water naturally "evaporates" into the air, but when we boil it, evaporation is more rapid. What exactly is taking place and does this extra water in the air have anything to do with humidity? A third question might be *how are evaporation and humidity related?* Finally, as you watch a pot of water boil you see "steam" rising into the air (some of you may even be familiar with the "steam" rising from geysers and hot springs). This looks vaguely like fog or a cloud. So a final question might be *what exactly is this "steam" and is it the same as a cloud?*

Since each of these smaller pieces builds on the previous one, it is important for you to have a solid understanding of each one before moving on to the next. This is important, and if you feel as though you do not quite understand something, talk to your instructor. You may find it useful to refer back to Figure C-1 on occasion, to see where we are and where we are going. This is particularly useful if you are having difficulty with a particular concept and you find yourself getting frustrated and bogged down on small details.

1	***EXPERIMENTING AND HYPOTHESIZING***

We begin this unit by looking at something we are all familiar with—thermometers. As you know, thermometers are used to determine the temperature of an object. What you may not be aware of is that thermometers do not always give a correct reading for the temperature of an object. Thus, we will begin by learning how to make an accurate measurement of temperature with a thermometer. After that, we will explore how the temperature of a glass of water changes when it is mixed with water of a different temperature. In particular, we will devise a method to determine the final temperature of the mixture after we combine a cup of hot water with a cup of cold water.

You may need some of the following equipment for the activities in this section:

- Thermometers and temperature sensors [1.1, 1.2]
- ~500 ml beakers [1.1, 1.2]
- ~100 ml graduated cylinders [1.1, 1.2]
- Styrofoam cups [1.1, 1.2]
- Hot and cold water [1.1, 1.2]
- MBL system [1.1, 1.2]

1.1   THERMOMETERS AND TEMPERATURES

Why does it take so long to take your temperature with an old-fashioned thermometer? What determines how long you must wait before the thermometer returns an accurate reading? You will be using thermometers in most of the activities in this unit. Therefore it makes sense to spend a few minutes investigating how they work.

Activity 1.1.1 Thermometers and Thermal Equilibrium

a)   Use a "standard" thermometer to measure the temperature of the air in the room. Next, get a cup of hot water from the tap and while watching the temperature readings on the thermometer, place it in the cup of water. Describe what you observe.

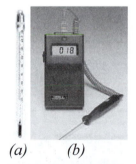

(a)   (b)

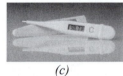

(c)

Figure C-2: Two types of thermometers: *(a)* A standard liquid-filled thermometer *(b)* and *(c)* Electronic-thermometers. (*(a)* and *(c)* courtesy of Pocket Nurse Enterprises, Inc. *(b)* courtesy of Electronic Temperature Instruments, Ltd.)

b)   Is the thermometer making an accurate reading of the temperature of the water from the moment it is placed in the cup? How long do you think the thermometer needs to sit in the water before it is reading an accurate temperature?

c) If the thermometer is not making an accurate reading of the water when it is first placed in the water, what temperature do you think it is measuring?

d) Now take the thermometer out of the water and let it sit in the room air. Again, watch the temperature readings and describe what you observe. How long do you think the thermometer needs to sit in the air before it is reading an accurate temperature? Why do you think this is so different from the time it needs to sit in the water?

There is a subtle lesson to be learned here. Before the thermometer is placed in the water, it is at the same temperature as the air. Once it is placed in the water, its temperature increases because it is in contact with the warmer water. Once it has been sitting in the water for a while, the thermometer reaches the temperature of the water. We say that it is in *thermal equilibrium* with the water. Thus, we view this situation as follows. When two objects at different temperatures are in thermal contact with each other, their temperatures will change. This will continue until the objects have reached thermal equilibrium, at which point their temperatures are equal. As simple and obvious as this may sound, scientists use this type of reasoning to define the concept of temperature.

Using a Temperature Sensor

The previous activity can be made much more visual with the use of an MBL system and a temperature sensor. The computer can make many temperature readings and plot them on a graph. This will allow us to view how the temperature is changing with time. This kind of graph is called a temperature-time graph. A temperature-time graph can be much more useful than a large table of numbers because you can see very quickly how the temperature varies over time. Your instructor may provide you with more specific information on how to use the temperature sensor and MBL system.

Activity 1.1.2 Temperature Sensors

a) Imagine the following experiment. A temperature sensor has been sitting out in room air for a while. The sensor is then placed in a cup of hot water until it is in thermal equilibrium with the water. Then the temperature sensor is removed from the water and held in the air. Make a rough sketch of what you think the temperature-time graph will look like for this experiment. Briefly explain your prediction. **Note:** Make sure to indicate when the sensor is placed into the water and when it is removed from the water.

Figure C-3: Temperature sensor for use with computer based laboratory systems. (Courtesy Vernier Software and Technology)

b) Now, carry out the experiment. Set the software to run for 10 minutes and start the experiment after the sensor has been sitting in the air for quite a while. Then, after about 30 seconds, place the sensor in a cup of hot water. After the temperature no longer appears to be changing, take the sensor out of the water and let it sit in the air for the remainder of the experiment. Print out a copy of your temperature-time graph to include in your Activity Guide and label the portions when the sensor is in the water and when it is in the air.

c) Is the behavior of the temperature sensor similar to that of the standard thermometer? Explain briefly

d) Describe how the graph looks when the temperature is changing rapidly compared to when it is changing more slowly.

It should be fairly obvious from the previous activity that thermal equilibrium is reached much more quickly when the temperature sensor is in water compared to when it is in air. It should also be pretty clear that the electronic temperature sensor behaves very much like a standard thermometer.

Accuracy of the Temperature Sensor

The fact that an electronic temperature sensor behaves much like a standard thermometer does not mean that it should be trusted without a second thought. As we saw in Activity 1.1.1, even a standard thermometer will give an inaccurate reading if it is not in thermal equilibrium with the object whose temperature you are trying to measure. Furthermore, there is some inherent imprecision when making *any* measurement (not just temperature). In the next activity, we will look more closely at how precisely we can measure temperature with our temperature sensors.

Activity 1.1.3 What Temperature is it?

a) Take a cup of cool water and place the temperature sensor in the water. After about 30 seconds, start the software while continuing to hold the temperature sensor in the water. Stop the software after 30 seconds and comment on the temperature-time graph.

b) Now, rescale your temperature graph so that the temperature range is about 1°C. For example, if you measured a temperature of 23.9°C, try scaling the graph to go from 23.5°C to 24.5°C°. Then try scaling it to an even smaller ranger say 0.5°C or 0.2°C. Comment on what you observe.

c) Based on your observations in part b), how much can you trust your temperature reading? Is it precise to within 1°C? To within 0.5°C? To within 0.1°C? Explain your reasoning.

As demonstrated in the previous activity, your temperature measurements will not be perfectly accurate. There will be some (we hope small) amount of uncertainty associated with any measurement process and it is wise to always keep this in mind. Knowing how much uncertainty is inherent in your measurements will help you determine when two measurements are the same. For example, if your temperature sensor is precise to within 0.5°C, and you measure the temperature of two cups of water to be $T_1=22.3°C$ and $T_2=23.6°C$, you can be pretty sure that these temperatures are, in fact, different. If, on the other hand, you measure these temperatures to be, $T_1=22.3°C$ and $T_2=22.9°C$, you cannot know for sure whether or not these two temperatures are different because they could each be wrong by as much as 0.5°C.

Most electronic temperature sensors are precise to within 0.5°C and many are precise to within 0.1°C (or even better). In this unit, we will not be interested in measuring temperature differences that are smaller than 1°C.

1.2   MIXING WATER

The next two activities constitute a game. The object of the game is to be able to predict the final temperature when two cups of water (initially at different temperatures) are mixed together. As you might guess, this is fairly easy if the two cups have the same amount of water, but it is not so easy when they don't. Before beginning an experiment it is useful to make a prediction about what you expect to see. It is important to make the prediction as specific as possible. For example, in the next activity you will predict the final temperature that results from mixing together a cup of hot water and a cup of cold water. You might predict that the final temperature will be somewhere in between the two starting temperatures, and you'd be correct. But a better prediction will be more specific about the final temperature. Will the final temperature be exactly halfway between the two temperatures? One third of the way?

Developing such a prediction involves considering what factors are likely to affect the experiment (e.g., the amount of water) and what factor are unlikely to affect the experiment (e.g., day of the week). Then the experiment tests your hypothesis. If the results of the experiment are inconsistent with your hypothesis, you know that it is false. On the other hand, if your experimental results are consistent with your hypothesis, your hypothesis is supported. It is important to remember that a hypothesis can never be proven correct. Even though a hypothesis may be supported by many observations and experiments, only one careful experiment is required to show that it is incomplete or incorrect it.

Activity 1.2.1 Final Temperature—Predictions

a) Suppose you mix two cups of water that are at different temperatures, one at a high temperature, T_H, and one at a lower temperature, T_C. What do you think the final temperature of the mixture will be? (i.e., will it be higher than T_H, lower than T_C, right in the middle, or something else?) Explain briefly

b) What factors do you think will influence the final temperature of the mixture? Explain.

c) To make your ideas a bit more concrete, predict the final temperature of a mixture of 50 grams of water[7] at 60°C mixed with 100 grams of water at 20°C. Try to explain the reasoning behind your prediction as best you can. **Note:** If you have to make a bit of a guess at this point, that's fine.

Testing Your Predictions

In this next activity you'll have a chance to test the predictions you just made. Remember that the goal is to come up with a method of predicting the final temperature when two *arbitrary* amounts of water at different temperatures are mixed together. So be prepared to make a number of different measurements.

---

[7] A gram is measurement of *mass*, which is a measure of "how much stuff" you have. One milliliter (ml) of water has a mass of 1 gram (gm)

Activity 1.2.2 Final Temperature—Observations

a) Typically, when beginning a scientific investigation, it is best to start simple. Therefore, you should begin by mixing together two cups of *equal* amounts of water (say, 50 or 100 grams each) that are at different temperatures. It might be difficult to get specific temperatures, so just use hot tap water and cool tap water. To record your observations, place an electronic temperature sensor in each cup and begin recording data. Then, after waiting until the sensors are recording a fairly steady value, pour the cold water into the cup with the hot water and place both temperature sensors in the final mixture. Print out a copy of your data to include in your activity guide. Is the final temperature halfway between the two initial temperatures?

b) You should have noticed that both temperature sensors end up at the same final temperature (within a few tenths of a degree Celsius). If they differ from each other by more than 0.5°C, ask your instructor how to calibrate your sensors. You should also have noticed that the final temperature was not exactly halfway between the two initial temperatures. Most often, the final temperature will be a bit below the halfway point (usually less than 1°C). Explain why this happens. **Hint:** What happens to the temperature of a cup of hat water that sits in a room for a long time?

c)   The fact that the final temperature was (probably) not exactly halfway between the two initial temperatures is worth thinking about briefly. There could be some outside influence that affects the experiments (as alluded to above), or there could be other measurement errors that lead to the unexpected result. Besides temperature measurements, what other measurements did you make in this experiment? How might errors in these measurements lead to a different final temperature than you expected?

d)   Now try the same experiment again, except, this time, combine 100 grams of cold water and only 50 grams of hot water. Again, write down the initial and final temperature and print out a copy of your graph to include in your activity guide. **Note:** Remember to pour the cold water into the cup of hot water.

e)   Using the data from this experiment, work with your group to determine an equation that allows you to calculate the final temperature you measured above. **Hint:** What fraction of the total mixture was initial in the cold water cup and what fraction was initially in the hot water cup? Do these fractions have any bearing on where the final temperature will lie between the two initial temperatures? Remember that your final temperature measurement might be a little lower than expected.

f)   When you have an equation that works, generalize it so that it is valid for arbitrary amounts of water and arbitrary initial temperatures. Show your work and write your new equation below.

g)   Using your newfound equation calculate the final temperature when mixing 50 grams of water at 60°C and 100 grams of water at 20°C. Do you results agree with your predictions from Activity 1.2.1? (It's okay if they don't)

h)   Try one more experiment with 100 grams of cold water and 25 grams of hot water. Compare your experimental measurements with your theoretical predictions. They should be quite close!

Congratulations! You have successfully developed a method for predicting the final temperature when mixing two cups of water together. This kind of quantitative reasoning plays an important part in the scientific process. It is equally important to consider sources of error and outside influences that may have affected your experiment. Each measurement that is made is a potential source of error so that making careful measurements is essential. Furthermore, even if perfect measurements could be made, outside influences (such as the cooling of the hot water) can also have an impact on the final results of an experiment. Often it is only possible to understand a particular result after including the outside influences and sources of error.

Checkpoint Discussion: Before proceeding, discuss your ideas with your instructor.

2	*THE DISTINCTION BETWEEN HEAT AND TEMPERATURE*

While most people are reasonably good at distinguishing hot from cold, our bodies are actually not very trustworthy when it comes to trying to determine temperature. In fact, it is not too difficult to fool yourself into thinking an object is hotter or colder than it really is. This happens, for example, if you place one hand in a tub of hot water and the other hand in a tub of cold water and leave them there for a little while. At this point, most people will be able to correctly distinguish which water is hotter. However, if you now take your hands out and put them both into water of the same temperature, one hand will feel hotter than the other. The hand that was in the hot water will sense the water as being colder than the hand that was in the cold water. If you didn't know better, you would incorrectly predict that the temperatures were *not* the same. This effect is quite dramatic and if you have never experienced it before, you should try it!

One of the things we will be considering in this section is exactly what we mean by how "hot" something is. Since we cannot trust our senses, we would like to develop a method of assigning a value for the "hotness" of something that is independent of our body's senses. Clearly, a thermometer is what we seek. However, although they are quite commonplace, many people do not understand exactly why a thermometer works the way it does. To gain an understanding of the inner workings of a thermometer, you will have an opportunity to build your own.

You may need some of the following equipment for the activities in this section:

- Liquid soap and ground black pepper [2.1]
- Styrofoam (or other insulated) cups [2.1 - 2.4]
- Hot and cold water [2.1- 2.3]
- Brownian motion demonstration (or use video clip) [2.1]
- Thermal expansion of a solid demonstration [2.2]
- Blocks made of metal, Styrofoam and wood with a hole drilled in them to accept the tip of a temperature sensor [2.2]
- Glass syringes and small flasks [2.2]
- Rubber tubing, one-holed stoppers, connectors [2.2]
- "Standard" thermometers [2.2]
- Thermos (preferably vacuum insulated) [2.3]
- An MBL system [2.3, 2.4]
- Temperature sensors [2.3, 2.4]
- Heat pulser [2.3, 2.4]
- Immersion Heater [2.3, 2.4]
- Anti-freeze [2.4]

2.1 GASES, LIQUIDS, AND SOLIDS

Figure C-4: Individual iron atoms are arranged in different patterns on a copper substrate and imaged with a scanning tunneling microscope. (Courtesy of International Business Machines Corporation)

Much of our knowledge about gases, liquids, and solids come from delicate experiments that are not easily reproduced in the classroom. One reason for this is the extremely small size of the constituents of matter. For example, the approximate size of an atom is about 0.0000000001 meters (10^{-10} m) in diameter. That means it would take roughly 100 million (10^8) atoms, side by side, to reach a length of one centimeter. Even more dramatic is that it would take about 10^{24} atoms, all clumped together, to occupy a cubic centimeter (a region about the size of a board-game die). That's a million billion billion, an almost incomprehensibly large number. Even the best optical microscopes in the world do not allow us to see things that small with our eyes. It is only with the use of exotic instruments, such as electron microscopes, atomic-force microscopes, or scanning-tunneling microscopes, that we can image individual atoms (see Figure C-4).

Before continuing, it will be useful to give a brief description of atoms and molecules. An atom is a stable configuration of three different tiny particles—electrons, neutrons, and protons. These particles are arranged in such a way as to behave like a single particle. There are many different configurations of electrons, neutrons, and protons that will form stable entities, and each one has its own unique properties. These entities are referred to as chemical *elements* and are listed in the periodic table, which groups them by weight and the similarity of their properties. A *molecule*, on the other hand, is a stable grouping of two or more atoms that behave as a single entity and has its own unique properties that might be very different from the elements that make it up. All of the materials you come into contact with everyday are made of large numbers of atoms or molecules, the properties of which are determined by the individual atoms or molecules.

As a simple example, oxygen is an element that most people are familiar with. However, what most people don't realize is that the oxygen that we breathe is actually *molecular* oxygen, consisting of two oxygen atoms bound together. We denote elemental oxygen with the symbol O and molecular oxygen with the symbol O_2. The subscript 2 refers to the fact that there are two oxygen atoms contained in this molecule. Similarly, three oxygen atoms can be combined into a stable configuration called ozone, with the symbol O_3. Furthermore, even though there are only about 100 elements, some molecules (such as DNA) can contain thousands, or even millions of atoms. While not every combination of atoms will form a stable entity, there seems to be no limit to the number of molecules that can be created. Inventing new materials that have special properties is an active area of research called *materials science*.

We won't worry too much about the distinction between atoms and molecules. For our purposes, what is important is the fact that they are extremely small, stable particles that usually are not deformed or altered. With this in mind, let us now loosely define the three most common phases of matter: solids, liquids, and gases.

Gases—A gas is a substance in which all the particles (atoms or molecules) have (basically) no attraction to each other. We can think of a gas as a collection of little hard spheres, that are whizzing around in random directions at very high speeds (about 1,000 mph). When two of these particles run into each other, they collide and bounce off one another, sort of like a collision between billiard balls. The important point is that there is no interaction between the individual particles unless they smash into each other. Another important point about gases is that because the particles move so fast and collide so often, the effects of gravity are almost completely negligible when discussing the individual gas particles. Thus, when a gas is placed into a container, the gas molecules quickly fly around and completely fill whatever container they are in.

Liquids—A liquid is a substance in which the particles (atoms or molecules) have an attraction for one another. (If the liquid is made up of molecules, such as water, then it is important to note that the attraction between the atoms in the molecules is *much* stronger than the attraction between the molecules.) This means that the individual water molecules want to be next to each other. Because of their attraction to each other, liquid particles behave "en-mass," with billions upon billions of individual particles acting together. Thus, unlike gases, gravity has a large effect on liquids. In addition, because the attraction between liquid particles is not all that strong, individual particles can move between other particles, allowing the liquid to "flow." Thus, when a liquid is placed inside a container the particles fall to the bottom and conform to the shape of the container.

Solids—A solid is a substance in which there are very strong attractions between the individual particles (atoms or molecules). (The strength of this attraction is less than the attraction between atoms in a molecule, but it is a much stronger one than the attraction between liquid particles.) In fact, this attraction is so strong, that the individual particles cannot move around like they can in a liquid. In some sense, you can think of a solid as an extremely large molecule, whose constituents may be either atoms or molecules. Now,

(a)

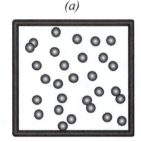

(b)

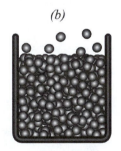

(c)

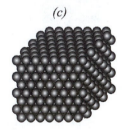

Figure C-5: Depictions of atomic models for (*a*) a gas, (*b*) a liquid, and (*c*) a solid.

although the individual particles are fixed to their neighbors, they can still wiggle around a little. It is as if the particles are attached to each other by little springs, which allow them to bounce around a little without moving away from each other.[8] Because of the strong bonds between neighboring atoms, a solid retains its shape regardless of what kind of container it is placed in (it doesn't flow like a liquid).

Measuring the Size of a Molecule

It might seem as if molecules are too small to be investigated, let alone have any importance in our day to day lives. We will soon see that thinking of substances as composed of atoms and molecules allows us to develop a very powerful theory for heat and temperature. And while it is true that molecules cannot be seen directly, it is possible to design experiments that do reveal information about lone molecules. In fact, using only a drop of soap, some water, and some pepper you can determine the approximate size of a single soap molecule!

Activity 2.1.1 Estimating the Size of a Molecules

a)   Take a small cup of water and gently sprinkle some pepper on the surface. Next, pull a hair out of your head and lightly dip one end in some liquid soap. You should end up with a small bead of soap on the end of the hair. Your bead of soap should look like a tiny sphere with a diameter about twice the width of your hair (see Figure C-6). Now poke this end of the hair into the center of the cup of water and describe what you observe.

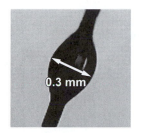

Figure C-6: A human hair with a small drop of liquid soap on it as seen under a microscope.

b)   What do you think happened to the drop of soap?

---

[8] In fact, the bonds between atoms in a molecule also behave like little springs.

c) You should have seen that an approximately circular area has been "cleared" of pepper. Estimate the radius of this circle and calculate the area in square millimeters. Recall that the area of a circle is $A = \pi R_c^2$, where R_c is the radius of the circle. If you measure the radius in millimeters, your area will be in square millimeters.

d) So what happened in this experiment? The small bead of soap that we began with spread out into a thin pancake when it hit the surface of the water. That is what pushed all the pepper out of the way. Basically, the soap tried to make itself as thin as possible. Clearly, the pancake of soap cannot be thinner than one molecular layer (although there's no guarantee that it isn't more than one molecule thick). Thus, in order to estimate the size of a soap molecule, we need to equate the volume of soap that we started with (the small sphere) to the volume of soap that we ended with (the thin pancake). This is shown schematically below. **Note:** It might help to visualize this using modeling clay. Roll some clay into a ball. This is like the soap droplet on the hair. Then squash the ball with your hand until it is a thin pancake. Notice that the shape of the clay has changed dramatically, but the amount (volume) of clay has remained the same.

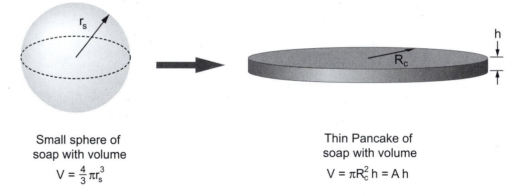

Small sphere of
soap with volume

$$V = \frac{4}{3}\pi r_s^3$$

Thin Pancake of
soap with volume

$$V = \pi R_c^2 h = A\,h$$

Estimate the volume of soap that was initially on the hair. (Remember that this is an estimate. You don't have to know the *exact* size of the sphere of soap, just the *approximate* size). **Hint:** The width of a human hair is about 0.1 mm.

e) Now equate this volume to the volume of a small pancake, whose area A, you estimated in part c), and solve for the height h, of the pancake This is your estimate for the smallest thickness that a single soap molecule can have.

Keep in mind that the size of the molecule you just calculated is only an *estimate* for the size of a soap molecule. In fact, since it is possible that the thin layer of soap is more than one molecule thick, all we can say for sure is that the size of a soap molecule can be *no larger* than the value you calculated in the last activity. However, this estimate should make it clear that the size of a soap molecule is many thousands of times smaller than the width of a human hair. Furthermore, since soap molecules are composed of many atoms, the size of an atom is smaller still.

Seeing Molecular Motion

Because the small size of atoms and molecules prohibits us from seeing them individually, we can't just look at them to understand how they behave. However, because molecular motion is important in understanding much of what is to come, we will attempt to give you some direct evidence for the motion of molecules. The following activity re-creates an observation that was first made in 1827 by the English botanist Robert Brown. Brown was looking through a microscope at pollen grains and other small inanimate objects suspended in water and was surprised to see them moving. This is now referred to as *Brownian motion*.

Activity 2.1.2 Brownian Motion

a) If your instructor has a Brownian motion demonstration, be sure to take a look. Keep in mind that the particles you are seeing are thousands of times *larger* than the water molecules themselves. It is also important to notice that the drop of water you are looking at appears to be completely motionless, like a glass of water that has been sitting for a long time. Describe the motion you observe as best you can.

Figure C-7: What do you see when you look at particles suspended in water under a microscope? Rather than moving smoothly, the particles jiggle around erratically. What do you think causes the particles to move in this way?

b) Does the motion of these particles tell you anything about the motion of the water molecules? Explain why or why not.

It is important to note that the water molecules are much smaller than the particles you are observing. What you are seeing when the particle moves is the effect of thousands of water molecules all hitting the big particle. An analogy would be many people together pushing a very large "Earth ball" or medicine ball. One person alone cannot move the ball. However, if many people push at various places on the ball then the ball might move in a certain direction. That direction will change, however, as the people move around in order to find room to push.

2.2   TURNING UP THE HEAT

Now that we have some evidence for the natural motion of molecules in a liquid, we can try to observe some differences between hot and cold objects. This is not as simple as it sounds. For example, imagine walking into a room where there are two glasses of water that look identical but they are at different temperatures. Without measuring their temperatures or feeling them in any way, would you be able to determine which one is hotter? Thus, we need to answer the question, "Is anything measurably different between these two glasses of water besides their temperature?" This is the subject of the following activity.

Activity 2.2.1 Heating Objects

a) Besides temperature, do you think there will be any measurable change in a solid if it is heated? If so, explain briefly what you think will happen.

b) Your instructor may have a demonstration of a solid object being heated. Watch carefully and describe what happens to the object when it is heated.

c) Explain how this behavior might be exploited to make a solid thermometer.

d) Explain how this same behavior in a liquid could be used to construct a liquid thermometer.

e) At the beginning of this activity, you saw that when a solid is heated, it expands. In terms of the actual molecules that make up a solid (or liquid), can you think of two possible explanations that would account this observation?

One very reasonable hypothesis that explains the expansion phenomenon (called *thermal expansion*) observed in the previous activity is that the molecules themselves increase in size when heated. However, there are some materials that actually contract when heated. Even more confusing is that certain materials expand when heated at some temperatures and contract when heated at other temperatures. For example, liquid water contracts when heated between 0°C and 4°C, but expands when heated above 4°C. Thus, although the idea that the molecules get larger when heated is appealing, it is difficult to use this idea to explain much of what is actually observed.

Molecular Motion and Temperature

An alternative idea that seems quite reasonable to most people is that the molecules gain energy and mover faster when heated. As the molecules jiggle around more and more, it seems plausible that they might "demand" a little more space. In fact, it is more complicated than this because the intermolecular forces can be such that the molecules actually "demand" *less* space when moving faster. While the details of this are beyond the scope of this class, numerous experiments have confirmed that when a material is heated, the sizes of the molecules do not change, whereas the average speed of the molecules increases (regardless of whether the material expands or contracts).[9] Therefore, in this class, we will consider *temperature* to be a measure of the underlying molecular motion of the material.[10]

Checkpoint Discussion: Before proceeding, discuss your ideas with your instructor.

The Need for a Thermometer

At the beginning of this section, it was mentioned that our bodies are not necessarily very good at determining the temperature of objects. The following activity explores this point a little further.

Activity 2.2.2 Which is Hotter?

a) You should have at your table a block of Styrofoam, a block of wood, and a block of aluminum. Briefly touch each of the objects

---

[9] In fact, in some situations, it is possible for molecules to gain energy without increasing their average speed.
[10] A more detailed study of temperature would reveal that this definition is somewhat limited. However, for our purposes it is more than sufficient.

and rank them in order from coolest to warmest. **Note:** Don't handle the blocks too long or you will warm them up with your hands.

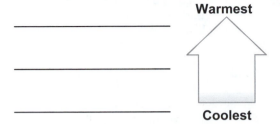

Warmest

_____

_____

_____

Coolest

b) Do you think these objects have the same or different temperature? If you think they are the same, explain why. If you think they are different, indicate which has the highest temperature and which has the lowest temperature and estimate the temperature difference between them in °C.

c) Now, measure the temperatures of each object by inserting a temperature sensor into the hole in each object. Note: Remember not to handle the objects too much because the heat from your hands will change their temperature.

Block Material	Temperature (°C)
Metal	
Styrofoam	
Wood	

d) Do your results surprise you? Explain briefly.

The last activity should make it very clear that what we feel as hot and cold does not necessarily coincide with the temperature of the object. That is, our bodies are not good thermometers, So how does one go about building a good thermometer? Well, we saw in Activity 2.2.1, that some solids expand when heated. You proposed a method for building a thermometer based upon this property common to many solids and liquids. Unfortunately, since solids and liquids undergo only a very small expansion when heated, it is not very easy to construct this type of thermometer in the classroom. Instead, we will build a constant pressure *gas* thermometer. But first, we need to consider the behavior of gases when heated.

Do Gases Expand?

Because it is the intermolecular forces that determine whether a material will expand or contract when heated, you might wonder what will happen to a gas when it is heated. Since we have described a gas as having no intermolecular forces, it may not be clear what should happen with a gas. The following activity examines this question.

Activity 2.2.3 Heating a Gas

a)   Recall that a gas expands to fill the container it is placed in. Image taking a fixed amount of gas and placing it into a large glass container with a lid that sealed the container and was immovable. Now, imaging heating up the gas, assuming that the size of the container cannot change. Would the gas expand? Explain briefly.

b)   Now if the temperature of the gas has increased due to the heating, will this have any effect on the collisions the gas molecules make with the walls? Explain briefly. **Hint:** What is temperature a measure of?

It should be clear that unlike a liquid or a solid, a gas does not have a well defined volume. Instead, the volume of a gas depends entirely on the size of the container it is placed in. Thus, a gas will expand or contract only if the volume of the container changes, independent of whether it has been heated or not. Now, if the container is flexible and has the ability to shrink or grow, then heating the gas will result in the gas pushing harder on the container which will cause the volume of the container to increase. This is the principle of the constant pressure gas thermometer.

Activity 2.2.4 Building a Thermometer

a)   To build the thermometer, place a one-holed rubber stopper in the mouth of a small flask or other container. Next, attach a piece of rubber tubing to the hole in the rubber stopper with connector or adapter. For the last step, you will need to choose a glass syringe that is approximately 20-25% of the size of the container that you are using.[11] Thus, if you are using a 125-ml flask, a 20-30 ml syringe works well. Finally, connect the free end of the rubber tubing to the glass syringe, making sure that the piston on the syringe is only pulled out approximately 25% of the way (if you are using a 20-ml syringe, the barrel should be pulled out to about 5-ml). A schematic diagram of a finished thermometer is shown in Figure C-8 below. Check to see that your thermometer works by placing the flask (not the syringe) in hot water and then in cold water. Make sure that the piston on the syringe is always pointed in the "up" direction.

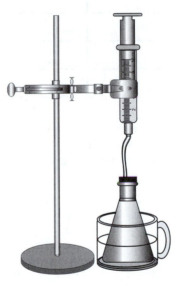

Figure C-8: A thermometer built from a flask and glass syringe sits in a cup of water.

---

[11] Using a syringe that is too small will result in the piston of the syringe being pushed all the way out of the barrel, which often leads to broken glass.

b) Now, the numbers on the side of your syringe correspond to actual temperatures, but you first need to calibrate your thermometer to a universally accepted temperature scale. Do this by placing your thermometer into a hot water bath of known temperature and labeling the position of the piston with this temperature using masking tape. Then do the same thing for a cold water bath of known temperature. Using these two markings, you should be able to determine what temperatures correspond to the other major markings on your syringe. Record these in the table below.

Syringe labels (ml)	Temperature (ºC)

c) Determine the maximum and minimum temperatures you can measure with your thermometer. Would your thermometer be useful for measuring very high or very low temperatures? Explain.

Although the thermometer you just built works reasonably well, it has a couple of major drawbacks. One is that this thermometer can only measure a limited range of temperatures. More problematic is that a change in atmospheric pressure will change the reading on the thermometer, even if the temperature does not change. This is one reason why people do not use such crude thermometers. However, this type of thermometer demonstrates something fundamental about all thermometers. They take advantage of a property of a material that changes with temperature. In this case, the gas causes the container to expand as its temperature is increased, and that expansion can be measured and calibrated. Although the expansion property of materials is a convenient feature to exploit in making a thermometer, it is by no means the only one. Other types of thermometers may rely on other properties of substances. The temperature sensors we have been using illustrate one common example. These sensors exploit the electrical properties of materials to produce an electrical current that changes with temperature. This current is then calibrated and displayed as a temperature on the computer screen.

2.3   THERMAL ENERGY TRANSFER AND TEMPERATURE CHANGES

Armed with our understanding of the concept of temperature, we will now turn our attention to trying to understand "heat." Clearly, we can increase the temperature of an object by heating it up, but what exactly does "heating it up," mean? One way of heating something up is to place it on a hot burner. But are there other ways? That is the topic of the next activity.

Heating Water with Mechanical Work

If temperature is a measure of how quickly molecules are moving, then perhaps we can increase an object's temperature by physically making the molecules move faster. In fact, pioneering experiments in heat and temperature tried to do just this. Scientists would spin paddle wheels in liquids and try to measure a change in temperature. You will do this in the next activity, albeit in a less controlled manner.

Figure C-9: A thermos is an insulating container designed to prevent heat from entering or leaving. (Corbis Images)

Activity 2.3.1 Shake it, Baby!

a)   Your instructor will place a small amount of water (50-100 g) in a thermos, which is an insulated container designed to prevent heat from entering or leaving the container. Write down the initial temperature of the water that is placed in the thermos.

b)   After the thermos is sealed, your instructor will begin shaking it vigorously. This is difficult work, so after about a minute or so, it will be passed to one of the students in class. This student should continue to shake the thermos vigorously for about a minute or so, then pass it on to someone else. When the thermos has gone around the whole room, pass it back to the instructor, who will open it and measure the temperature of the water inside. (Since this will take some time, your instructor might want you to work ahead on the next activity while all the shaking is going on.) Write down the final temperature of the water after having been shaken.

c) From a molecular perspective, what is different about the water after having been shaken and what caused the change? (You should think a bit more deeply than just replying, "the shaking caused the change.")

Heating Water with Electric Work

This following activity involves heating up some water while monitoring its temperature in a way that might be a bit more familiar. You will be using an immersion heater, which is similar to a stove's burner. When on, electricity causes the immersion heater to get hot and become a "source of heat." We use the immersion heater instead of a stove because it is easier to control. It is also a more delicate piece of equipment and you need to be careful with it.

Caution: *Never plug the heat pulser into the wall when it is not in the water.* This will damage the heating unit. It is all right to plug the heat pulser into the computer, but make sure it is immersed in water before actually pulsing heat. Another thing to remember is that when the heater is on, *the metal portion of the heating unit shouldn't touch anything except the water*. This includes the air in the room, you, and your partner!

Activity 2.3.2 Pulsing Heat in a Cup of Water

a) Begin by putting 100 grams of water into a Styrofoam cup and placing a heat pulser in the water with a stand. Set the total experiment time to 250 seconds and set the heat pulse length to 10 seconds. When you think you are ready, have your instructor check your set-up *before* starting the experiment! You will need to stir the water with the temperature sensor throughout this experiment, being careful not to touch the heat pulser with the temperature sensor. Begin the experiment and after about 40 seconds (don't forget to stir the water continually), click on the heat button. This will add a 10-second pulse of heat, which should increase the temperature of the water a little bit. After the temperature of the water has stabilized (which should take about 40 seconds) hit the heat button again. Repeat this process until the experiment ends. When you are done, you should have hit the heat button five times, once at 40, 80, 120, 160, and 200 seconds. Briefly describe the graph you obtained, and print out a copy for your activity guide.

b) Fill in the following table by finding the temperature change for each pulse of heat that was added.

Heat Pulse	Temp. before pulse (°C)	Temp. after pulse (°C)	Temp. Change (°C)
#1			
#2			
#3			
#4			
#5			

c) Explain what a 10-second pulse of heat does to your 100 grams of water. **Hint:** Is the temperature change for each pulse approximately the same?

d) How much would you expect the temperature of 50 grams of water to rise when subjected to one 10-second heat pulse? What about 10 grams of water?

e)   From a molecular perspective, what is different about the water
     before and after receiving a pulse of "heat?"

Heat as an Energy Transfer Process

As these last two activities showed, you can increase the temperature of water by placing
a "heat source" in the water (the heat pulser), but you can also increase the temperature of
water just by shaking it (with no heat source at all). This surprises many people.
Remember, an increase in temperature just means that the molecules are moving more
vigorously, so anyway that you can get the molecules moving more vigorously will result
in an increase in temperature. By shaking the water, you are adding energy to the water
which speeds up the motion of the molecules. In scientific terms, *work* is being done on
the water.

Notice that whether we used an electrical source to "heat up" the water or we did work to
"heat up" the water, the underlying result is the same in both cases; there is an increase in
the motion of the water molecules. Viewed from this perspective, it may be clearer to
think in terms of the "energy transferred" to the water. This *thermal energy* either comes
from collisions with the molecules that make up the heating unit or from the energy that
you used to shake up the thermos.

Quantifying Heat

Now, since the result is the same in both cases (i.e., the temperature of the water
increased), we might choose to define "heat" (or *thermal energy*) in terms of the
temperature change of water. In fact, this is precisely how scientists first defined heat:
One *calorie* is the amount of thermal energy needed to raise the temperature of *one gram*
of water by *one degree Celsius*. **Note:** The term *calorie* (with a small c) as defined here is
not the same as the term *Calorie* (with a capital C) in regards to food consumption. One
"food" *Calorie* is actually equal to one thousand *calories* as defined here. In this activity
guide, unless specifically stated, we will always be referring to "non-food" *calories*. The
following activity should help you get familiar with this unit of thermal energy.

Activity 2.3.3 Counting Calories

a)   If it takes one calorie to raise the temperature of one gram of water
     by one degree Celsius, how many calories does it take to raise the
     temperature of 100 grams of water by one degree Celsius?

b) How many calories would it take to raise the temperature of 100 grams of water by 5 degrees Celsius?

c) How much thermal energy (in calories) is required to change 50 grams of water from 30 degrees Celsius up to 50 degrees Celsius?

d) How much thermal energy did each pulse of your heat pulser produce in the Activity 2.3.2? Show the details of your calculation.

2.4   SPECIFIC HEAT

Because we have defined a calorie in terms of its effect on one gram of water, it is particularly easy to determine how much thermal energy has been transferred to a specific amount of water if we know its temperature change. However, there are many situations when one might be interested in heating up a substance besides water. So the question naturally arises, if we transfer one calorie of thermal energy to one gram of a substance that is not water, will its temperature also increase by one degree Celsius? That is the topic of the following activity.

Activity 2.4.1 Activity: Pulsing Heat Again

a) Begin by placing 100 grams of anti-freeze into a Styrofoam cup and placing an immersion heater in it. Set the total experiment time to 250 seconds and prepare to deliver 10-second heat pulses to the water. **Note:** When you think you are ready, have your instructor check your set-up *before* starting the experiment! You will need to continually stir the anti-freeze with the temperature sensor throughout this experiment, being careful not to touch the heat pulser with the temperature sensor. Begin the experiment and after about 40 seconds, click on the heat button. After the temperature of the anti-freeze has stabilized (which should take about 40 seconds) hit the heat button again. Repeat this process until the experiment ends. When you are done, you will have clicked on the heat button five times, beginning at 40, 80, 120, 160, and 200 seconds. Briefly describe the graph you obtained, and print out a copy for your activity guide. Is there any difference between this graph and the graph you made in Activity 2.3.2?

b) Fill in the following table by finding the temperature change for each pulse of heat that was added.

Heat Pulse	Temp. before pulse (ºC)	Temp. after pulse (ºC)	Temp. Change (ºC)
#1			
#2			
#3			
#4			
#5			

c) Knowing how much thermal energy is transferred to the anti-freeze in one 10-second pulse (see Activity 2.3.2), determine how much thermal energy it takes to increase the temperature of one gram of anti-freeze by one degree Celsius. Show your work.

––––––––––––––

The quantity just calculated—how much thermal energy it takes to change the temperature of one gram of anti-freeze by one degree Celsius—is an important and useful quantity. It is a property of the material that does not change as you add more material. Scientists call this quantity the *specific heat* of a substance and denote it with the symbol c. We already know that it takes one calorie to raise the temperature of one gram of water by one degree Celsius. This means that water has a specific heat of $c = 1 \text{ cal}/\text{g}^\circ\text{C}$. Since the specific heat tells us how much energy it takes to increase the temperature of *one* gram of a substance by one degree Celsius, the quantity mc (where m is the mass in grams) is the amount of energy it takes to increase the temperature of m grams by one degree Celsius.

Stated another way, $mc\Delta T$ is the amount of thermal energy gained by an object of mass m when its temperature changes from T_i to T_f. If $T_f > T_i$, then $\Delta T > 0$ and the energy *gained* by the object is positive. But if $T_f < T_i$, then $\Delta T < 0$ and the energy gained by the object is negative. This indicates that the object actually *lost* energy. Now, when two objects are placed in thermal contact with each other, the energy *gained* by one object will be exactly equal to the energy *lost* by the other object. We can use this fact to determine the final temperature of two cups of water when they are mixed together.

Activity 2.4.2 Thermal Energy Transfer

a) Suppose we have m_h grams of hot water, initially at a temperature of T_h and m_c grams of cold water initially at temperature T_c. After mixing, the final temperature is given by T_f. Write down an expression in terms of m_c, c, T_c and T_f for the heat *gained* by the cold water during the mixing process?

b) Write down an expression in terms of m_h, c, T_h and T_f for the heat *lost* by the warmer water during the mixing process. **Hint:** The quantity will be positive.

c) Now, equate the heat *gained* by the cooler water to the heat *lost* by the warmer water and solve for the final temperature T_f.

d) Compare your equation to the equation you deduced in Activity 1.2.2 and show that these two equations are identical.

Checkpoint Discussion: Before proceeding, discuss your ideas with your instructor.

The equation you derived here should be the same as the equation you deduced in Activity 1.2.2 from your experimental observations. Hopefully, you can appreciate this result even more now, knowing that it comes from equating the energy *lost* by the hot water to the energy *gained* by the cold water. In fact, this is actually one of the most important principles in all of the sciences. It is better known as the principle of energy conservation. Although we have not discussed it in detail, energy comes in many different forms, motion, thermal (heat), sound, chemical, solar, electrical, nuclear, etc. The conservation of energy principle states that energy can never be created or destroyed, it can only be changed from one form to another, or transferred from one object to another. We will not try to confirm this statement, but it is worth mentioning that it has been well verified experimentally and is one of the most fundamental ideas in the all the sciences.

3	EVAPORATION, CONDENSATION, AND HUMIDITY

At this point, you should have a pretty good understanding of temperature and thermal energy transfer. In this section, we will be taking a look at some situations where a transfer of energy does *not* result in a temperature change. This will lead us to consider how substances can change from one phase into another. In particular, water is a substance that readily changes from a liquid into a gas. In this section we will investigate how this phenomenon can be understood in the context of energy transfer. Specifically we will try to answer the question: is the transfer of energy necessary for a substance to change phase?

You may need some of the following equipment for the activities in this section:

- Styrofoam (or other insulated) cups [3.1 – 3.2]
- Temperature sensors [3.1 – 3.3]
- Immersion heater [3.1]
- Heat pulsers (optional) [3.1]
- MBL system [3.1 – 3.3]
- Mass balance [3.1]
- Hot and cold water [3.1 – 3.3]
- Paper towels [3.2]
- Relative humidity sensors [3.3]
- Plastic, seal-able containers [3.3]

3.1   HEAT OF TRANSFORMATION

We saw in the last section that adding one calorie of thermal energy resulted in one gram of water changing temperature by one degree. Thermal energy is almost always discussed in the context of *transfer* from one object to another. Based on this observation, we might conclude that if we continue to add energy to water its temperature will continue to rise. Let's find out if this is true. We know that if we add thermal energy to a container of water it will eventually begin to boil. The question we want to ask is what happens to the temperature of the water if we continue to add energy to it while is boiling? **Caution:** Be very careful in the following activity. Boiling water can cause severe burns.

Activity 3.1.1 Watching Water Boil

a)   Begin by measuring the mass of an empty cup. Now put about 150 grams (it doesn't have to be exact) of room temperature water in the cup and measure the mass of the cup/water system. Report your results below.

Figure C-10: When water is heated to 100 °C, it begins to boil. What happens to the water when it boils? (Eye Ubiquitous /Corbis Images)

b) Set up the temperature sensing software so that it will take readings once per second for about 15 minutes and make sure the temperature range goes from 0°C to 110°C. Next, place the temperature sensor and immersion heater (without plugging it in) into the water. Have your instructor check out your setup **before** starting the software. Once you have been checked out, start the software and begin stirring the water with the temperature sensor, being careful not to touch it to the heating unit. After 20 seconds or so, plug in the heating unit and continue stirring (without splashing) until the water begins to boil vigorously. At this point, you can stop stirring the water, but you still need to make sure that the temperature sensor doesn't come into contact with the heating unit.

After the water has been boiling vigorously for at least five minutes, unplug the heating unit and remove the temperature sensor from the water. Then, as quickly as possible (but being as careful as possible), remove the heating unit from the cup and measure the mass of the cup/water system. Report your result below. *Do not throw out the water remaining in the cup!* If there is less water in your cup than before you boiled it, where do you think the water went?

c) After about 10 minutes (when the water has cooled off substantially) measure the mass of the cup/water system again. In the meantime, you should print out a copy of your data for your activity guide. How much water was lost while cooling down? Do you think any water was lost as the water was heating up (but not yet boiling)? Using this information, determine how much liquid water was lost during the boiling process alone. Show all you work.

d) Briefly describe the main features of your graph, specifically what happened to the temperature of the water once it started boiling. What does this mean about how quickly the water molecules are moving? Are they moving faster and faster as you add energy to the boiling liquid? Explain briefly.

e) Although you continued to add energy into the water, its temperature stopped increasing. Where did the energy that you added to the water go?

At this point, you might have noticed that our definition of "heat" is a bit flawed. We defined our unit of thermal energy (the calorie) as the amount of energy needed to increase the temperature of one gram of water by one degree Celsius. Thermal energy was studied by its affect on the temperature of a substance. As the previous activity just showed, the temperature of water did not increase beyond 100°C, even though we were still adding energy. Thus, adding one calorie of thermal energy to one gram of water that is already at 100°C will *not* increase its temperature by one degree, which makes it important to specify what temperature the water is at when we try to increase its temperature. The accepted definition for one calorie is the amount of thermal energy needed to increase the temperature of one gram of water from 14.5 °C to 15.5 °C.

The fact that the temperature of the water stopped rising at 100 degrees Celsius may not have been a real surprise to you. However, you may not have realized that this temperature actually tells you something about the interaction between the molecules of water. Since temperature is a measure of the average speed of the molecules, it seems like there will be a point when the water molecules are moving so vigorously that they can "break away" from each other. This is what's referred to as the *boiling point* of the material. At this point, the liquid molecules are breaking their bonds with each other and becoming gas molecules. The bubbles you see when boiling water are bubbles of *gaseous* water (not air), also called steam.[12]

This is an example of what scientists call a *phase change*; the state of matter is being changed from one phase into another. In this case, liquid water is being turned into gaseous water. This happens at a well-defined temperature called the *boiling point*.

---

[12] It is worth pointing out that when people say they see "steam" rising from a pot of boiling water, they are not actually seeing steam. We will discuss this in more detail when we talk about cloud formation.

Another example is when solid water (ice) is turned into liquid water. This happens at a well defined temperature called the *melting* (or freezing) point. Because the inter-molecular forces are different for different molecules, each substance has its own unique melting and boiling points.

Determining the Energy of a Phase Change

The fact that it takes some energy to "break apart" a solid into a liquid or a liquid into a gas should not be completely surprising. This energy is called a *latent heat,* although latent energy is probably a better term. The latent heat of substances is very important in many areas of science and engineering. With our data from the last activity, we can calculate the *latent heat of vaporization*, which tells you how much energy must be transferred to one gram of water to break the molecules apart and create steam. In the next activity you will determine the latent heat of vaporization of water.

Activity 3.1.2 Heat of Vaporization

a) Examine your graph of the boiling water again. Notice that once the temperature reaches about 60 ºC, the line starts to curve down a little bit (can you explain why?) Before that, it looks like a very straight line. Fit a straight line (either by hand or with the computer) to the portion of the graph that looks like a straight line (up to about 50 ºC). Write down the value for the slope of this line (including units) and interpret what this slope means physically. **Hint:** Remember that the slope is the "rise" over "run." What does the "rise" (the change in *y*) tell you *physically*? What does the "run" (the change in *x*) tell you *physically*?

b) Knowing how much water you had at the beginning of the experiment, use the slope of this line to determine how much energy is being transferred to the water every second (or minute). **Hint:** How many calories does it take to increase the temperature of your cup of water by 1°C and how many degrees Celsius is the temperature increasing every second (or minute)? Show your calculations!

c) Now that you know how much energy is being delivered to the water every second (or minute), determine how much total energy was delivered to the water during the time it was boiling. **Note:** You will need to determine from your graph exactly how long the water was boiling.

d) Finally, knowing how much energy was delivered during the boiling process and how much liquid water was turned into steam during the boiling process, you should be able to determine how much energy it takes to turn *one gram* of liquid water into steam. **Hint:** Remember that some water might have been turned into steam before or after the water was actually boiling.

e) This quantity is called the *latent heat of vaporization* of water, and is given the symbol L_v. The accepted value for water is 539 cal/g. Compare your result to this and comment on any difference?

Your measurements for the heat of vaporization of water should be reasonably good. There may be some small experimental errors, but if you are careful, you should get a result that is within 5% of the accepted value. It should be pointed out that there is an analogous quantity called the *latent heat of fusion*, which tells you how much energy it takes to *melt* one gram of a substance. Although we won't actually measure this quantity (it would make a nice project), the procedure would be similar to the experiment done here. As already mentioned, the latent heat of vaporization, L_v, tells you how much energy it takes to vaporize one gram of water. Thus the quantity mL_v tells you how much energy it will take to vaporize m grams of water. For example, if you have 10 grams of water, (10 g)(539 cal/g) = 5390 calories will be required to vaporize the water.

3.2   EVAPORATION

Most people are familiar with the fact that a small puddle of water on the ground will eventually disappear. This happens because of a process called evaporation. To begin our investigation of evaporation, we will first examine a quantity called the "wet-bulb" temperature.

Activity 3.2.1 What is Room Temperature?

a)   Suppose two electronic temperature sensors are sitting in a room (22°C). One is initially surrounded by a piece of paper towel that has been soaked in warm (30°C) water, the other is open to the air. Sketch below your prediction for the temperature of the two thermometers over ten or so minutes.

b)   Now try the experiment. Set up the software to run for at least ten minutes. Then, get a cup of water that is warmer than room temperature (at least 5°C warmer) and wrap the end of one of the temperature sensors with a small piece of paper towel and tape it securely in place. Place this sensor in the water and stir it around till the temperature stabilizes. When the temperature is stable and only about 5 degrees higher than the thermometer that's not in the water, start the program and take the thermometer out of the water and lay it down on the table (on top of some paper towels so the table doesn't get wet). When the program has finished running, print out a copy of your graph. Describe what happens. What is the lowest temperature reached by either thermometer?

c)   Which of the thermometers do you think gives a better indication of "room temperature?" Why.

d) Based on your observation, what would happen if a thermometer wrapped in a towel soaked in 22°C water were left open to 22°C air. Do you think the temperature would drop below 27°C? Discuss this with your group and come to a consensus. **Hint:** was the paper towel in the last experiment ever at 22°C?

e) Using your knowledge of temperature, what can you conclude about the motion of the molecules in the wet paper towel versus the room air?

"Wet" and "Dry" Temperatures

The temperatures you just took are referred to as "dry-bulb" and "wet-bulb" temperatures. You will have noticed that the wet-bulb temperature reading drops down below the dry-bulb temperature reading and then levels off to a constant value. It is this constant value that it called the wet-bulb temperature. Most students find it surprising that the wet-bulb temperature drops below the dry bulb. Why does this happen and what stops it from continuing to drop? The answer must have something to do with the wetness of the paper towel, since that is the only difference between the two thermometers. The following thought experiment will help you understand why this happens.

Activity 3.2.2 Evaporating Water

a) Consider a glass of water that has no lid on it. If you wait for a really long time, the water in the glass will have *evaporated*. Where must the water molecules be going.

b) Since boiling water also disappears (although at a much faster rate), do you think these processes (boiling and evaporating) are related? Explain briefly.

c) What causes specific water molecules to leave the water? Can *any* water molecule leave or only certain ones? Explain. **Hint:** When water is boiling, lots of molecules are leaving the water.

d) Remember that temperature is actually a measure of molecular motion. If the wet-bulb reading is lower than the dry-bulb reading, what does that tell you about the average motion of the molecules? Use this to explain how evaporation could be the cause of the wet-bulb reading being lower than the dry-bulb reading.

Checkpoint Discussion: Before proceeding, discuss your ideas with your instructor.

We have developed the idea that temperature is a measure of the average speed of the molecules. This means that the lower wet-bulb reading must be a result of slower moving molecules. However, the wet-bulb reading was originally at a higher temperature, which is a result of faster moving molecules. Clearly the average speed of the molecules around the wet-bulb thermometer is decreasing. Now, one way for the faster moving (water) molecules to slow down, is though collisions with the slower moving (air) molecules. The wet paper towel and the air are in *thermal contact* with each other, so at some point they should reach *thermal equilibrium* (i.e., they would reach the same temperature). However, this is not what we observed. Something *else* must be happening to cool the wet-bulb reading below the dry-bulb reading. This "something" is called evaporative cooling and is an example of a dynamic equilibrium (as opposed to thermal equilibrium).

Evaporative Cooling

Evaporation occurs when molecules leave the bulk liquid and move off into the air. As we saw when boiling water, the ability of molecules to leave the liquid is enhanced when the liquid temperature is greater. This should make sense, since faster moving molecules (and temperature is a measure of molecular speed) are more likely to have the necessary energy to break away from surrounding molecules. But the temperature of a liquid is only the measure of the *average* speed of the molecules. This means that some molecules are moving *faster* than average, while others are moving more slowly. Which are more likely to break free from the liquid? In fact, only the fastest moving molecules are in a position to break away. You observed this earlier when boiling the liquid. If all of the molecules could break free, then the entire cup of water would suddenly vaporize into steam. This in fact, is not what happened. It took some time for the water to boil away. So what happens as the faster moving molecules leave? The *average* speed goes down. Since temperature is a measure of this *average* speed, the temperature goes down as well. This is called *evaporative cooling*, since the evaporation process itself lowers the temperature.

Since heating the water hastens evaporation, one might wonder if qualities of the air (other than temperature) also affect evaporation. The answer, as you will soon discover, is yes. A clue is found in how we cool ourselves during exercise. Sweating uses the idea of evaporative cooling to lower your body's temperature. If you are from New Orleans or another humid city, you might know from first-hand experience that this doesn't always work. Sometimes sweating only makes you wetter. In particular, on a very humid day, sweating is a very inefficient means of cooling off. We'll see why in the next section.

3.3   RELATIVE HUMIDITY AND DYNAMIC EQUILIBRIUM

It is well known that placing a lid on a glass of water will prevent it from evaporating. The act of covering the water prevents molecules from leaving the glass. This seems reasonable enough, however, the cover is not in contact with the surface of the water so how can it possibly prevent molecules from leaving the water? Shouldn't the water continue to evaporate? This question is the addressed in the following activity.

Activity 3.3.1 Dynamic Equilibrium

a) If the cover on the glass of water cannot prevent molecules from leaving the water, what do you think happens to the molecules that *do* leave the water? Where do they go?

b) We observe that in a sealed container the water level does not change. This means that the total number of molecules in the liquid is not changing. However, some water molecules are leaving the liquid (the fastest moving ones). How can you reconcile these two seemingly contradictory facts? **Hint:** What must be happing in order for the number of molecules in the liquid to remain constant?

c) If the total number of molecules in the liquid is not changing, yet some molecules are *leaving* the liquid, then it must be the case that some of the gaseous water molecules are *returning* to the liquid. In fact, if the total number of molecules in the liquid is not changing, then the number that leave must be equal to the number that return. If more molecules leave the liquid than return, then the water will slowly disappear. If we know how many molecules are leaving and returning to the liquid each second, then we can determine how the number of gaseous water molecules changes as shown schematically in the figure below.

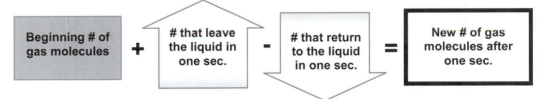

Now, imagine a glass of water that is sealed with a lid. Imagine further that at the instant the lid is placed on the glass, there are no water molecules in the gaseous state (they are all still in the liquid). After a short time, some of the liquid molecules have evaporated and become gaseous water molecules above the liquid. Since the surface of the liquid water does not change its size, it is reasonable to assume that the same number of water molecules leave the liquid every second. Let's assume for the moment that 100 molecules leave the liquid every second.[13] We need to devise a model for how the molecules might return to the liquid. We could assume that this process is similar to evaporation. That is, every second a certain number of molecules return to the liquid, regardless of other factors. On the other hand, we could assume that the number of molecules that return to the liquid depends on the number of gaseous water molecules above the liquid. Discuss with your group which of these two ideas seems more realistic. Explain your reasoning below.

---

[13] In fact, the number of molecules evaporating each second depends critically on how much of the water is exposed to the air. Still, a more realistic estimate of this number would be one hundred million billion or 10^{17} molecules every second.

d) Most students find it more realistic to assume that when there are a small number of gaseous water molecules, there will be a small number returning and when there are a large number of gaseous water molecules, there will be a large number returning. A simple way to model this is to postulate that a certain percentage of the gaseous water molecules will return to the liquid each second. For simplicity, let's assume that 50% of the gaseous molecules return every second. Using these assumptions, we are now in a position to determine the number of gaseous molecules as a function of time. This calculation has been started in the table below. Complete this table, rounding your results to the nearest whole number. Note that the number of gas molecules one second later (right column) becomes the beginning number of gas molecules for the next second.

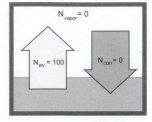

t = 0 seconds

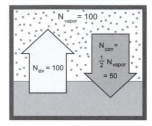

t = 1 seconds

Time (sec)	Beginning # of gas molecules	# leaving (constant)	New # of gaseous molecules one second later
		# returning (50% of # at left)	
0	0	100 / 0	100
1	100	100 / 50	150
2	150	100 / 75	175
3	175	100 /	
4			
5			
6			
7			
8			
9			
10			
11			
12			

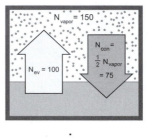

t = 2 seconds

t = 12 seconds

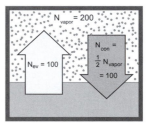

Figure C-11: If molecules leave (evaporate) the liquid at a constant rate N_{ev} each second and return (condense) to the liquid at a rate N_{con} per second proportional to the number of molecules in the gaseous phase, the system will eventually reach a dynamic equilibrium in which the total number of molecules in the gas (vapor) phase (N_{vapor}) is constant.

e)   Describe what happens after about 10 minutes, and using the data from your table, make a rough sketch of the number of water molecules in the gaseous state as a function of time.

Gaseous water molecules are also referred to as *water vapor*. Thus, the graph you just made depicts the amount of water vapor in a sealed container as a function of time. Initially, the amount of water vapor increases quite rapidly, but then levels off to a constant value. This happens when the *evaporation rate* (number of molecules leaving the liquid each second) is exactly equal to the *condensation rate* (number of molecules returning to the liquid each second). This is what scientists call a *dynamic equilibrium*. This simply means that although the amount of water in the glass does not appear to be changing, it is actually losing and gaining molecules at exactly the same rate so that the average number of molecules in the liquid does not change. Although the mathematical model we used in the previous activity is quite simple, it captures the main features of evaporation in a closed container. In reality, the actual percentage of molecules evaporating and returning both depend on many factors, such as temperature and the amount of air above the liquid.

When the rate of evaporation is equal to the rate of condensation, we say that the air is *saturated*. This simply means that no more water molecules can coexist with the air. If more water molecules are placed into the air, they simply *condense* back into liquid water. Thus, there is no *net* evaporation. This is why a glass of water that is covered will not disappear. After a little time, there is enough water vapor coexisting with the air above the liquid so that an equal amount is being condensed back into the water as is being evaporated away. The amount of water vapor that can co-exist with the air at saturation is called the *equilibrium value* and must be experimentally determined. It has been observed that the equilibrium value for water vapor in air increases as the temperature increases.

When the amount of water vapor is far from its equilibrium value, the rate of evaporation is much larger than the rate of condensation and there is a net evaporation. But as the amount of water vapor in the air gets closer and closer to its equilibrium value, the number of molecules that condense back into the liquid state increases and this slows down the net rate of evaporation. Thus, things evaporate more quickly when there is a small amount of water vapor in the air and things evaporate more slowly when there is a large amount of water vapor in the air.

Evaporation, Humidity, and "Wet" and "Dry" Temperatures

Remember that this investigation started with the observation that a thermometer surrounded with a damp towel read a *lower* temperature than a dry thermometer. This led to the idea of evaporative cooling and the factors that affected evaporation. We now return to our wet and dry thermometers for a thought experiment.

Activity 3.3.2 Water Vapor Indicator

a) We saw earlier that wet-bulb and dry-bulb temperature measurements are not equal. Would you expect a higher temperature difference between these readings when there is more or less water vapor in the air? Explain.

b) How do you think the wet-bulb and dry-bulb readings would compare with each other if the air were saturated? Explain.

c) Explain how you could use wet-bulb and dry-bulb temperature readings to tell you something about the amount of water vapor in the air.

The difference between wet-bulb and dry-bulb temperatures cannot tell you exactly how much water vapor is co-existing with the air. It can only tell you how much water vapor there is relative to the amount at saturation. This is an important point because as already mentioned the saturation value depends on temperature. Thus, the wet-bulb and dry-bulb readings will be the same whenever the air is saturated, even though the actual amount of water vapor in the air might be very different. This relative measurement of water vapor can be used to define a quantity called the *relative humidity*. Relative humidity is defined as the amount of water vapor in the air divided by the amount of water vapor in the air at saturation (the equilibrium value). Thus, relative humidity tells you, on a percentage basis, how close the water vapor content of the air is to it's equilibrium value.

Because the amount of water vapor at saturation depends on temperature, relative humidity itself depends on temperature. In fact, this temperature dependence plays an important role in the formation of clouds, as we will see shortly. Unfortunately, it makes the determination of relative humidity a little bit more complicated because the wet-bulb and dry-bulb temperature difference (sometimes just called the wet-bulb depression) will change depending on both temperature and relative humidity! However, the meaning of relative humidity is always the same. A reading of 50% relative humidity means the water vapor content of the air is half its equilibrium value. A 90% relative humidity reading means that the water vapor content of the air is 90% of its equilibrium value.

The following graph shows how you can use the wet-bulb and dry-bulb temperatures to determine the relative humidity. Notice that the same wet-dry temperature difference can correspond to different values of relative humidity depending on the (dry) temperature of the air. Again, this is because the equilibrium value of water vapor in the air depends on temperature. You can determine the relative humidity for values not shown on the graph by estimating where they would lie on the graph.

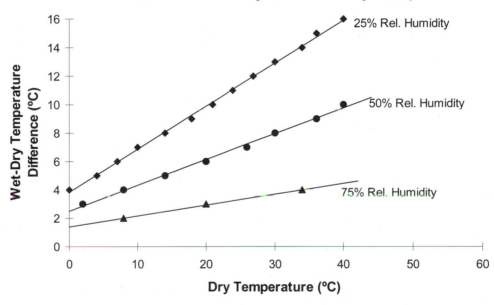

Using an Electronic Sensor to Measure the Relative Humidity

In addition to wet-bulb and dry-bulb temperature readings, relative humidity can also be measured using electronic components that undergo a physical change depending on the humidity (the same way our temperature sensors use electronic components that undergo physical changes depending on the temperature). The following activity will have you measuring the relative humidity of a sealed container. When using the relative humidity sensor, be careful to keep it out of water.

Activity 3.3.3 Measuring Relative Humidity

a)   Place the relative humidity sensor on a small stand inside a large plastic container (a small cooler works well). You may want to tape the sensor to the stand so that it doesn't fall off into the water. Now set the experiment length to 15 minutes and use a data rate of 1-2 points per second. Before pouring any water into the container, begin the experiment and let it run for a minute or two so that you have a good reading of the relative humidity of the room. Then, pour in a cup or two of hot tap water (enough to completely cover the bottom of the container) and seal the container. Describe below what you expect will happen to the relative humidity readings as time goes on. Make a sketch of what a graph of relative humidity versus time would look like.

b)   When the experiment ends, print out a copy of your data for your activity guide. Does the relative humidity reading reach 100%? Explain why this makes sense. If it doesn't reach 100%, can you explain why?

c)   Compare your graph to the prediction you made as a result of your dynamic equilibrium model from Activity 3.3.1. Do the graphs have the same shape? Explain why you might expect these two graphs to look similar.

Checkpoint Discussion: Before proceeding, discuss your ideas with your instructor.

The last few activities have introduced a number of new terms and concepts such as relative humidity, saturation, wet-bulb depression, equilibrium value, etc. It is beneficial to discuss these terms and concepts with your group to make sure that you understand what they refer to. We will be using these terms in the next section when we discuss cloud formation.

4	*CLOUD FORMATION*

Have you ever wondered exactly what a cloud is? Many people are surprised to find out that clouds are nothing more than a collection of tiny droplets of water (or possibly ice crystals). Indeed, looking up into the sky and seeing huge white billowy clouds often makes one think more of cotton than droplets of water. On the other hand, you know from experience that it only rains when there are clouds present, so maybe it is not all that surprising that clouds are nothing more than water droplets. In this section, we will be taking a closer look at how clouds form. By the time we are finished, you should understand exactly what ingredients are necessary for clouds to form, and you will get a chance to produce your very own cloud.

You may need some of the following equipment for the activities in this section:

- Petri-dish, small stand, and salt [4.2]
- Styrofoam cup [4.2]
- Fire syringe demo [4.2]
- Small can of compressed air [4.2]
- Small mass (~50 g) [4.2]
- Flask with one hole stopper [4.3]
- Rubber tubing and connectors [4.3]
- Small hand pump [4.3]

4.1   GAS COOLING AND DEW POINT

In order to understand cloud formation, it is essential that you know how the relative humidity changes with temperature. You should recall that the equilibrium value of water vapor in air increases with temperature.

Activity 4.1.1 Humidity and Cooling

a) Using the definition of relative humidity, explain why the relative humidity decreases when the temperature rises, even if no water vapor is added to the air.

b) Now imagine a parcel of air that is being cooled with no change in its water vapor content. In other words, the air is getting cooler but the number of evaporated water molecules in the air is unchanged. Explain how its humidity will change.

c) If the temperature of this parcel of air continues to drop, describe what happens to the relative amounts of evaporation and condensation.

Relative Humidity and Clouds

It is important to remember that the relative humidity does not tell us absolutely how much water vapor is in the air. Instead, it indicates the ratio of the amount of water vapor in the air to the maximum amount that *could* be in the air. The fact that the equilibrium value of water vapor in cooler air is smaller means that the relative humidity of the air will go up as the air is cooled. The graph below shows the equilibrium value of water vapor as a function of temperature. **Note:** Recall that after Activity 3.3.1, we defined the *equilibrium value* of water vapor in the air to be the amount of water vapor that will co-exist with the air once the system has reached equilibrium (i.e., the evaporation and condensation rates have reached the same value)

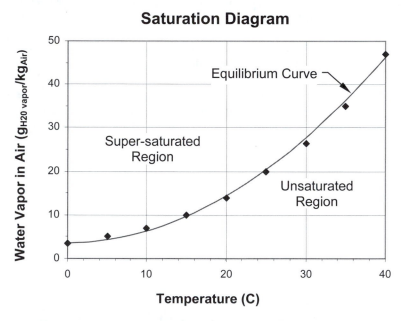

Figure C-12: A saturation diagram depicts the equilibrium curve for water-vapor. That is, the amount of water vapor that will co-exist with the air at equilibrium, as a function of temperature. The region below the curve includes systems that are unsaturated (the evaporation rate is larger than the condensation rate). The region above the curve includes systems that are saturated (the evaporation rate is smaller than the condensation rate). Systems lying on the equilibrium curve are referred to as saturated because the evaporation rate and the condensation rate are equal.

To get this *equilibrium curve*, we have plotted the maximum possible water vapor content of air at different temperatures. Note that the water vapor content is measured in units of grams of water per kilogram of air, so a measurement of 5 g/kg means that there are 5 grams of water vapor in every kilogram of air. Unlike the relative humidity, this is an *absolute* measurement of how much water vapor is in the air. The equilibrium curve represents the line of 100% relative humidity.

Consider a system (of air and water vapor) at a temperature of 15°C with 10 grams of water for every kilogram of air. Looking at the graph, you see that this air, represented by a point at 15°C and 10 g/kg, is sitting on the equilibrium curve. This means the water vapor is at its equilibrium value, i.e., the system is *saturated* and has a relative humidity of 100%. If this system suddenly warmed to 30°C without gaining any extra water, there would still be 10g of water vapor for every kg of air. This point, at 30°C and 10g/kg, lies below the equilibrium curve, which has a value of 27.5 g/kg at 30°C. This means that another 17.5 grams of water can co-exist with each kilogram of air (at this temperature). The region below the equilibrium curve is called the *unsaturated* region, since an air-water system in this region is not fully saturated. The relative humidity for this system at 30°C is calculated by dividing the actual water vapor content of the system by the equilibrium value at 30°C. That is,

$$\frac{10g/kg}{27.5g/kg} = 0.364 = 36.4\%.$$

Note: Strictly speaking we should always refer to "the air-water vapor system" when talking about relative humidity and saturation. In practice, however, it is often more succinct to refer to this system simply as *the air*. Thus, for convenience, we will often refer to a mixture of air and water vapor as a *parcel of air*.

The above calculation raises an interesting question. What if a parcel of saturated air gets colder? The next few activities explore this question.

Activity 4.1.2 Calculating Relative Humidity Changes

a) If you started with air that had a relative humidity of 50% at 40 degrees Celsius and you cooled that air down to 30 degrees Celsius (without changing the water vapor content), determine how the relative humidity changes.

b) If you started with air that had a relative humidity of 20% and a temperature of 20 degrees Celsius, and you added (for example, by boiling water) 5 grams of water vapor to the air *without* changing its temperature, determine how the relative humidity would change.

c) The *dew point* is defined to be the temperature at which a given parcel of air becomes saturated. This is the temperature at which the air has a relative humidity of 100%, assuming no water vapor is gained or lost. What is the dew point if the temperature is 25°C and the relative humidity is 50%?

d) What is the dew point if the air is 30°C with a relative humidity of 30%?

e) What do you think happens to the water in the air as it cools below the dew point?

Checkpoint Discussion: Before proceeding, discuss your ideas with your instructor.

Supersaturation

We have defined the saturation curve as representing the maximum possible amount of water vapor that air can hold. This is in fact not completely correct. If you cool air very quickly then the water vapor does not have time to condense out of the air. This system is then said to be *supersaturated* and is represented by points in Figure C-12 above the saturation curve. In this state, however, the water vapor in the air is very *unstable*, meaning the slightest change causes condensation. When this happens, the water vapor often, but not always, condenses into very small water droplets that we see as fog or as a cloud.

4.2   UNDERSTANDING CLOUD FORMATION

There are many different kinds of clouds and many different ways for them to form. We will limit ourselves to two situations. The first is when the temperature of air suddenly changes, as can happen in the atmosphere when thermal energy is absorbed from the sun or when air moves to colder regions. The other situation of interest is when two different parcels of air come into contact with each other. We call this a *mixing* cloud because the two air masses mix together, and the properties of the combined air mass can be quite different than either of the constituent air masses.

What Factors Contribute to Cloud Formation?

It is interesting to note that a cloud does not always form when the air reaches saturation. Consider a closed bottle of water. If the water has been sitting for a long time, the relative humidity of the air above the liquid will be 100%. This means the system is saturated. Nevertheless, you don't often see little clouds in bottled water (you do however, see clouds when opening a bottle containing a carbonated beverage). Clearly, saturated air is only one ingredient in cloud formation. The water vapor that exists in a saturated (or supersaturated) system is ripe to condense into water droplets. The following activity introduces you to the essential ingredient that leads to the condensation of water in the form of a cloud.

Activity 4.2.1 Condensation Nuclei

a)   Take a small Petri-dish or other small container and place a little room temperature water in it. Next, place a small platform (the bottom ¼" or so of a Styrofoam cup works well) in the water so that there is a dry, flat surface above the surface of the water. Now, half of the groups should place about 10 grains of salt on the small platform and the other half should leave theirs empty. Lastly, half of the "salt" groups and half of the "non-salt" groups should cover their platforms with a Styrofoam cup (you may need to place a small mass on top of the cup to keep it from floating). Gently put this little contraption aside so that it won't get bumped, we will be checking back on it later

b)   After about 45 minutes, take a look at the platforms that were left uncovered. Describe what you observe. Make sure you check experiments that had salt and those that didn't. Is anything unusual occurring on the platforms?

c) Now remove the cups from the other experiments and look at the platforms. Describe what you observe. Again, make sure to observe those that had salt as well as those that didn't.

d) For those experiments that were covered, what do you think the relative humidity was inside the cup? **Hint:** Do you think there was any net evaporation in this situation? Do you have any ideas why there was condensation on the grains of salt but no where else? Explain.

As seen in the last activity, the presence of the salt particles helps *encourage* condensation, but only when the relative humidity was 100%. The salt particles (or any other particles that act in this way) are called *condensation nuclei*. We will see later that condensation nuclei can have a large impact on whether or not a cloud can form.

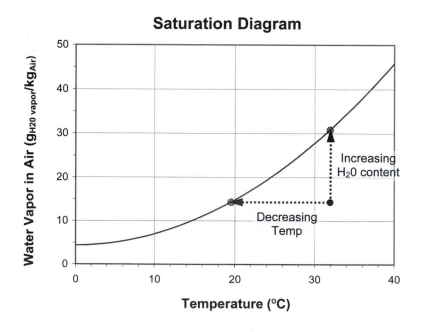

Now, before attempting to make a cloud, we need to understand how a system becomes saturated. There are two basic mechanisms by which a parcel of air can become saturated. One is by increasing the water vapor content of the system and the other is by decreasing the temperature of the system. These two processes are sketched on the saturation diagram above. Of course, real clouds usually result from a combination of these processes.

We will be making a cloud in a bottle by decreasing the temperature of the air. But first, we need to understand how to cool down a parcel of air. This is the topic of the next activity.

Activity 4.2.2 Expanding and Compressing Air

a) What do you think would happen if we were to rapidly compress some air? Would its temperature rise, fall, or stay the same? What makes you think so?

b) Now observe what happens to the temperature of a gas that is compressed. Take a well-sealed compression piston, known as a "fire syringe," and *very rapidly* compress the air in the tube. Describe what you observe (have your instructor help you with this). Can you explain this behavior using the concept of molecular motion?

c) What do you think would happen if you were to allow the air to rapidly expand instead of compressing it? Explain.

d) Now take a small can of compressed air and let some of the air escape. This air is rapidly expanding. Describe what you observe when you let some air out of the can. How does the can feel?

The results of the previous activity are pretty dramatic and may have surprised you a bit. Recall that in the beginning of this unit, we increased the temperature of water just by shaking it. That is, we did *work* on the water to increase its temperature. When compressing the air, we are similarly doing work on the air and likewise, its temperature increases. Of course, the temperature of the air increases a whole lot more than the water, but that's because there is so much less air than water. On the other hand, when the gas is escaping from a can of compressed air, the gas itself is doing work as it pushes air out of the can. For our purposes, we are only interested in the fact that when air is rapidly released from a container the temperature of the air left inside drops dramatically.

4.3   MAKING A CLOUD

We now have all the ingredients we need to actually create our very own cloud! This is the topic of the next activity.

Activity 4.3.1 A Cloud in a Bottle

a) Take a small flask and put about 10-20 grams of water in it. Next, connect a rubber tube from a small hand pump to the flask with a rubber stopper. You are now ready to try and make a cloud as follows: Pump some air into the flask with the hand pump and then let the flask sit for a moment. After letting the flask sit for a few minutes, what are the temperature and relative humidity of the gas inside the flask?

b) If you were to quickly allow the air to escape, how would the temperature of the air in the flask change? What would this do to the relative humidity? Would you expect to see a cloud?

c) After waiting for a minute to allow the gas to equilibrate, quickly pull the stopper out of the flask. Did you see a cloud? If you didn't, can you think of any reason why not?

Figure C-13: With an understanding of how temperature and humidity interact, it is possible to make a cloud in a bottle.

d) This first attempt at making a cloud may have been a dismal failure. The reason is that there may not have been enough condensation nuclei around. Try the experiment again, only this time, drop a lit match into the flask (with water) before putting on the rubber stopper. This will put many smoke particles in the flask to act as condensation nuclei. Once again, pump some air into the flask and then let it sit for a brief time. Explain what happens this time as you allow the air to expand suddenly.

Checkpoint Discussion: Before proceeding, discuss your ideas with your instructor.

Congratulations! You have just created your very own cloud. Now, in case this activity seems contrived to you, it is important to point out that this process is actually quite realistic. That's not to say that clouds actually form in bottles, but that air masses often cool down and become saturated.

Cloud Levels

Under normal conditions, the temperature of air decreases by about 2°C with every 1,000 ft. of altitude gained. Thus, as air rises it will cool. There are a number of reasons why air might rise. Wind may push an air mass against the side of a mountain forcing the air mass to rise. Two winds blowing in opposite directions might collide and force air upward. Whatever the reason, it is common for air to rise and if, as a result, the air cools to the dew point, there is a strong possibility that a cloud will form.

Activity 4.3.2 Cloud Levels

a) If the temperature at the surface of the Earth is 30°C and the relative humidity is 50%, at what altitude are you likely to find clouds? Explain. **Hint:** Air at the surface will probably get pushed upwards

Cloud Formation Due to Air Mixing

The final topic that we will undertake in this unit is that of a *mixing cloud*. The most common example of a mixing cloud is the cloud that forms from your breath on a cold winter day. This type of cloud forms when two masses of air mix together. The two masses of air might have different properties or they might be very similar. For example, one might be hot and humid and the other might be cold and dry. To understand the transformation of the air as it mixes together, it is instructive to consider carefully what happens to the temperature and the vapor content of the air.

Activity 4.3.3 Mixing Air Masses

a) Let us consider two parcels of air. The first one is at 20°C and has a water vapor content of 5 g/kg and the second one is at 30°C and has a water vapor content of 15 g/kg. Plot these two air masses as points on the following saturation diagram.

Saturation Diagram

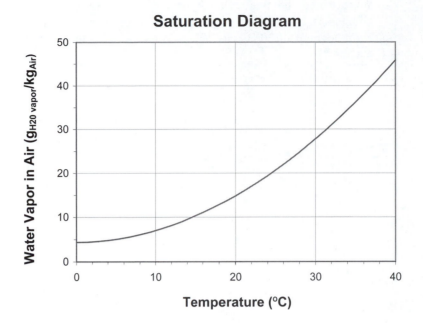

b) Now consider the case where both packets of air are the same size. For example, say both contain one kilogram of air. When they mix together, there will be two kilograms of air. What is the total amount of water vapor if these two packets combine? What is the water vapor content of the mixture in grams per kilogram of air?

c) Determining the final temperature of the new air mass is exactly the same as determining the final temperature after mixing two cups of water together. The final temperature will depend on the amounts and temperatures of the initial parcels of air. In this case, the initial amounts are the same, so the calculation is easy. Find the final temperature of the mass of air and, using the water vapor content from the previous question, plot the point that represents the final air mass on the saturation diagram. Describe where it lies in relation to the two initial points.

d) What would the final relative humidity be if, instead of one kilogram of each air packet, we had one kilogram of 30 °C air and two kilograms of 20 °C air? Where would it lie on the saturation diagram?

Notice that in the above activity, the final state after mixing lies directly between the initial points. The fact that it lies exactly in the middle is because we began with the same amount of air in the initial parcels. If we had started with one parcel being larger than the other, the final point would not be exactly in the middle. However, the final point would like *somewhere* on the line that joins the two initial points. In fact, it will be proportionally closer to the initial point that had more mass. Thus, if one parcel of air had three times the mass of the other, then the final point would lie three times closer to that point than the other, but still on the line that joins the two initial points.

Now consider the case where the two initial parcels of air are such that when you draw a line connecting them on a saturation diagram, that line goes into the super-saturated region. That suggests that there is the possibility of observing a cloud. Of course, it will depend on the initial masses of the air parcels and also on whether there are any condensation nuclei around. So, while there is no guarantee that a cloud will be visible, it is at least a possibility.

Activity 4.3.4 Mixing Clouds

a) Consider being outside on a cold winter day when the temperature is 2°C and the relative humidity is 50%. When you breath, you blow out air at that is near your body temperature of about 37°C with a high relative humidity of about 95%. Find these points on the following saturation diagram and connect them with a line. Do you think there will be a cloud on your breath? Explain your reasoning.

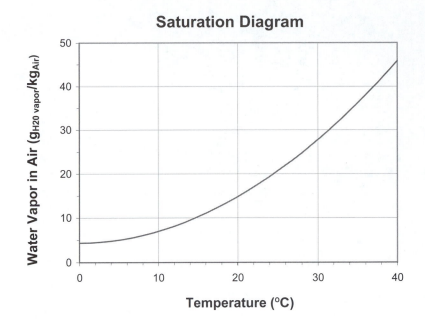

Saturation Diagram

Checkpoint Discussion: Before proceeding, discuss your ideas with your instructor.

It is surprising just how often mixing clouds will form. In fact, now that you understand where they come from, you will probably start noticing them all over the place. They happen at the exhaust pipes of cars, jet aircraft sometimes make them in the sky, you see them when you take a hot shower, and also when boiling water to make tea. The next time you see a mixing cloud, it may be difficult *not* to think about the water vapor content of the air masses and how their mixing leads to the air being in the super-saturated region of the saturation diagram.

5	*PROJECT IDEAS*

It is now time for you to take on the role of scientific investigator and to design a research project focused on some aspect of this unit that you found particularly interesting. On the pages that follow, you will find a number of project suggestions. Please do not feel limited by these suggestions. You may modify any of these or come up with a completely new one on your own. We have found that many of the best projects are those dreamt up by students. We therefore encourage you to develop your own project on a topic that you find interesting. You should of course consult with your instructor as some projects require too much time or impossibly large resources. Nevertheless, anything involving heat, temperature or humidity is fair game. So use your imagination and have fun!

Your instructor may ask you to write a brief proposal that outlines the goals of your project and how you plan to accomplish them. You may find it helpful to refer to the project proposal guidelines included in Appendix B. Try to plan your project in stages, so that if you run into difficulties early on you will at least be able to complete the data collection, analysis, and interpretation. To this end, it is important to note that the project proposals listed here are intended to foster your creativity, not to tell you exactly what to do. In most cases, answering all the questions in one of these proposals would take far more time than you have. So, choose a few questions that interest you or generate some of your own, but try to keep your project focused.

You will probably want to keep a lab notebook to document your project as it unfolds. Also keep in mind that you may be presenting your project to your classmates, so be prepared to discuss your results, how you measured them, and what conclusions you can draw from them. You may find it helpful to look over the oral presentations guidelines and project summary guidelines in Appendix B as you work. These guidelines may give you a better idea of what is expected from a typical student project. Be sure to consult with your instructor about their requirements for your project as they may differ from the guidelines laid out in Appendix B.

Good luck, and have fun!

5.1   SALTING ROADS AND MAKING ICE CREAM

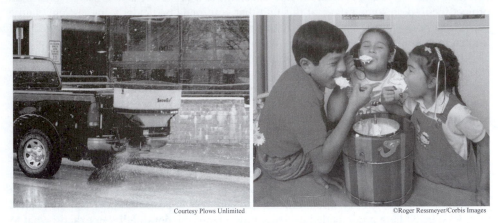

Courtesy Plows Unlimited   ©Roger Ressmeyer/Corbis Images

Figure C-14: Rock salt is used to melt snow on the roads. It is also used in the making of homemade ice cream. How does the addition of salt change the properties of water and ice?

Salting is one of the primary tools used to combat icy road conditions in the winter. In preparation for snow and ice storms, communities all over the world spread salt on the roads to melt ice. This works because the freezing temperature of salt water is lower than the freezing temperature of pure water.

It is also common to use salt when making homemade ice cream. Rock salt is added to ice in the outer chamber of the ice cream maker. But how does this rock salt help you make ice cream?

Your task is to develop an understanding of how salt and other impurities affects the freezing of water into ice. The following questions may provide some guidance as you explore the affects of salt on water.

1.   How does the addition of salt affect the freezing temperature of water? Does the freezing temperature change if more salt is added? What is the minimum freezing temperature that can be achieved by adding salt to water?

2.   Is it possible to change the freezing temperature of water by adding other substances (sugar, baking soda, alcohol, etc.). How does adding each substance affect the freezing temperature. What is the minimum freezing temperature you can achieve by adding each substance to water? Why do you think salt is most commonly used on icy roads?

3.   How does the temperature of ice change when salt is added? Start with crushed ice in an insulated container and add rock salt to it. Is the system exchanging thermal energy with the surroundings?

4.   How does the addition of salt or other impurities affect the temperature of water? Measure the temperature of a container of water as you add salt. What happens to the temperature? How does the addition of salt or other impurities affect the boiling temperature of water? What might be the benefit of adding salt to water when cooking? **Note:** These effects can be small so be very precise with your measurements

5.2   WHEN DOES COFFEE COOL THE FASTEST?

Imagine you and a friend have just visited your local coffee shop each picked up a cup to drink in class. As you go to put cream and sugar into your steamy cup of java your friend remarks, "I wouldn't do that if I were you. That's going to make your coffee cool down more quickly. If I were you I'd get take a couple of packets of cream and sugar and add them when we get to class." "That doesn't make any sense," you argue. "It shouldn't matter when I put in the cream and sugar." But your friend is adamant that adding cream and sugar earlier will make the coffee cool down faster.

After several minutes of arguing, you are both so sure of yourselves that you bet a months supply of coffee that adding cream and sugar earlier won't make the coffee cool down faster. In order to settle your bet, you decide to measure the affects of adding cream and sugar on the temperature of your coffee.

Design and carry out an experiment to determine if adding cream and sugar makes coffee cool faster. As you design your experiment think carefully about what outside factors might affect your results. You may also find it useful to compare several similar measurements in an effort to identify experimental errors. The following questions may be helpful as you explore how adding cream and sugar affects the temperature of coffee.

1.   What does the cooling curve (temperature as a function of time) look like for a cup of hot coffee? How long does it take the coffee to reach room temperature? Does this depend on the initial temperature of the coffee and/or the temperature of the room.

2.   How is the cooling curve changed if cream and/or sugar are added to the coffee when it is hot? How does the cooling curve change if cream and/or sugar are added to the coffee after it has cooled significantly. Does the amount of cream and/or sugar affect the cooling curve. Does stirring affect the cooling curve?

3.   Using an insulating container, you can thermally isolate the coffee from the room. In an insulated container does the addition of cream and/or sugar cause a significant change in the temperature of the coffee?

5.3   WINTER CLOTHING, INSULATION AND LAYERING

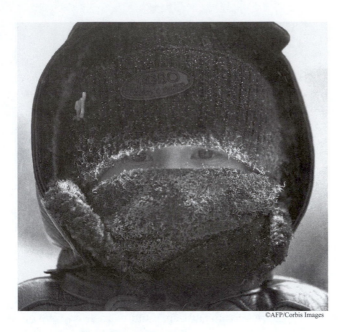

©AFP/Corbis Images

People use many types of clothing to protect themselves from the cold. We know that wearing clothes keeps us warm even when the air is very cold, but how do insulators keep us warm? Why do people wear different types of insulation in different conditions?

Without protective clothing most of us become uncomfortable at temperatures above about 80 degrees or below about 60 degrees Fahrenheit. None the less, people from all over the world find themselves exposed to extreme weather conditions (extreme cold, extreme heat, wind, rain, snow, humidity, etc.)In order to maintain a comfortable body temperature, they use many types of insulation from natural materials like down, fur, wool, and cotton to high-tech synthetic insulators like polyester fleece, Thinsulate™, and aluminized Mylar™ (the material out of which those handy little emergency blankets are made). With such a wide variety of insulating materials and weather conditions, how can one choose which insulation to use?

Pick several common insulating materials (for example: a cotton sweatshirt, a polyester fleece, a wool sweater, a rain jacket) and devise a method for investigating the insulating properties of these materials. A simple method might involve measuring the cooling curve (temperature vs. time) of a beaker of hot water wrapped in each insulating material.

The following suggestions/questions may provide some useful guidance as you investigate the insulating properties of common clothing materials.

1. Compare the cooling curves for similar beakers of hot water wrapped in different insulating materials. Which cool the fastest? Which cool the slowest?

2. Which is a more effective insulator, a single thick layer of material or several thinner layers adding up to the same amount of material. **Note:** Carefully consider how you will define the same amount of material (same weight, same thickness or what?).

3. How are the insulating properties of materials affected by moisture? To what extent do materials retain their ability to insulate when wet? Folklore among outdoor enthusiasts suggests that cotton is a very poor insulator when wet. To what extent is this true?

5.4   WIND CHILL, APPARENT TEMPERATURE, AND HYPOTHERMIA

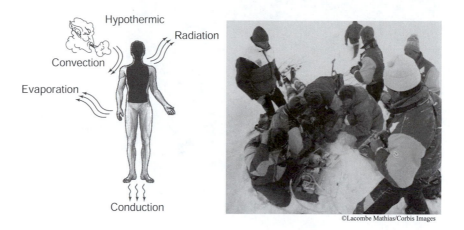

©Lacombe Mathias/Corbis Images

Under normal conditions, the human body maintains a constant temperature of 98.6 degrees Fahrenheit (37 degrees Celsius). Because our bodies hold a constant temperature, we can surmise that they have reached some sort of equilibrium with the environment. Our bodies maintain this equilibrium by adjusting the amount of thermal energy we produce (e.g. through physical exertion or shivering which triggers the release of thermal energy from chemical reactions in our muscles) and how quickly that energy is released to the environment (e.g. by controlling perspiration).

When exposed to certain environmental conditions, the human body is unable to maintain a constant temperature. If someone's body temperature drops below normal they are said to be hypothermic. A person who is hypothermic for too long or whose body temperature drops far below normal may die. Obviously, exposure to extremely cold temperatures can bring on hypothermia, but what other conditions can make a person become hypothermic?

Using a beaker of water and a heat pulser, design and build a simple thermodynamic model of the human body. Construct your model so that it can maintain a fairly constant temperature. Explore how different environmental conditions affect the temperature of your model person. The following questions may help you as you explore the affects of different environmental conditions on your thermodynamic model of the human body.

1. During the winter, it is common for the local weatherperson to report the wind chill in addition to the actual temperature. What does wind chill mean? We all know that it feels colder when it is windy than when it is not. But does the air temperature actually change? How does wind affect the ability of your model person to maintain its body temperature?

2. It is often argued that being wet can increase a persons risk of becoming hypothermic. Test this hypothesis with your model. Try the following: a) Model the situation in which a person is swimming in cool water. What happens if your beaker is partially submerged in a larger container of cool water? How does the water temperature in the outside bath affect the temperature of your model? b) Model the situation in which a person is exposed to rain. What happens if you wet the outside of the model and let the water evaporate? c) What if your model is exposed to wind and rain simultaneously?

5.5   EVAPORATION RATE

Figure C-15: The Hetch Hetchy reservoir formed by the O'Shaughnessy dam is a major source of fresh water and hydroelectric power for the San Francisco area. (©Galen Rowell/Corbis Images)

Much of the fresh water we drink is drawn from manmade reservoirs. These reservoirs are most often formed by damming up deep valleys. For example, the O'Shaughnessy dam which forms the Hetch Hetchy reservoir in California's Sierra Nevada Mountains was built in one of the deepest valleys in the region. One of the major reasons given for building reservoirs in deep valleys is that for an equivalent volume of water, deeper reservoirs experience less evaporation.

Your task is to explore the factors that affect evaporation from a body of water. Before you begin, write down a list of factors that might affect the evaporation rate from a body of water. Design an experiment in which you can control one or more of these factors

As you design your experiment, it may be helpful to look back at the activities in subsection 3.3. Here, you observed the relative humidity increase when you sealed a container. You probably deduced that this happens because as water evaporates the amount of water vapor (the number of water molecules) in the air increases. Thus, the rate of change of relative humidity (slope of the relative humidity curve) in a sealed container is directly related to the net evaporation rate. In addition to making experimental observations, you may want to use a procedure similar to the one from Activity 3.3.1 to develop a spreadsheet model of evaporation. Using your experimental data, you should be able to adjust the parameters of your spreadsheet model to match the behavior of the real world.

You may find the following questions helpful in guiding your investigation

1. Since evaporation involves the movement of water molecules from the liquid into the air, the amount of the water's surface exposed to the air might affect the evaporation rate. Devise an experiment to measure the evaporation rate for vessels containing the same amount of water with different surface areas. How does the evaporation rate change as the surface area is increased? How can your incorporate this behavior into your spreadsheet model?

2. More water vapor seems to rise from a cup of hot coffee than from a cup of cold coffee. This suggests that the evaporation rate may depend on the temperature of the liquid. How does the evaporation rate of the liquid change with the temperature of the liquid? How can you incorporate this into your spreadsheet model?

3. We know that the relative humidity in a closed container will always eventually reach 100%. Does this happen at the same rate for any container. How does the rate at which the relative humidity reach 100% change as the size of the container is changed? How can you incorporate this behavior into your model?

5.6   TEMPERATURE AND HUMIDITY VARIATIONS

©Michael and Patricia Fogden/Corbis Images

Throughout each day, the temperature and relative humidity are known to vary significantly. In the night temperatures typically drop and during the day they rise. Similar daily variations in humidity are easily observed. It is most common to see dew on plants in the morning. What drives these changes in temperature and humidity?

Many would argue that there is a clear relationship between temperature and humidity because "humidity decreases as temperature increases!" Again this is easily observed. As the temperature rises during the day, early morning dew and fog vanish. However, other people would argue that this relationship is only true if there is no change in the amount of water vapor in the air, and the amount of water vapor in the air can change throughout the day.

As a knowledgeable citizen with an interest in atmospheric phenomena, you and some friends decide to answer this question on your own. Devise a method to measure the temperature and humidity on a regular basis over several days. You can also calculate how the relative humidity would change if there were no change in the water vapor content of the air. Comparing this prediction with the actual experimental data will tell you how much the water vapor content changes during a typical day.

Some things to keep in mind:

1.   Does it matter how many times you check the temperature and humidity throughout the day? What do you think is a good number?

2.   Can you think of any reasons why the water vapor content should either change or remain the same throughout the day?

3.   Is it critical to use the same set-up for all of your measurements? Can you think of a way of designing a set-up that will not need to be changed at all for a two or three day period?

4.   As you look over your results note any significant changes in the water vapor content of the air. Do these correlate in any way to the weather?

5.7   THE DENSITY OF FOG

©Craig Aurness/Corbis Images

Dense fog is a common weather phenomenon in many parts of the world. Places like San Francisco and London are famous for having dense fog. Why is fog so common in these areas? While it is often beautiful, fog can also cause many problems for travelers. Driving a car or flying an airplane in dense fog is very dangerous because visibility can sometimes be dramatically reduced to only a few meters. In an effort to protect pilots and drivers from accidents, it is desirable to predict when heavy fog will form.

Your task is to measure how some factors affect the density of fog in order to give meteorologists a basis for predicting when heavy fog will form. In Activity 4.3.2, you made a cloud in a bottle by rapidly decreasing the pressure in a very humid environment. Using this as your model, investigate how different environmental parameters affect the density of the fog. For example you might vary the temperature of the water in the flask or the amount of air you pump into the flask.

You will need to devise a way to measure the density of fog. A simple method for doing this might be to shine a light though the clear flask containing your "cloud in a bottle" and use a light sensor to detect the amount of light that reaches the other side. As the fog becomes more dense, the amount of light that penetrates the cloud will decrease.

Use the following questions as a guide in your investigation.

1. Using the "cloud in a bottle" experiment as a starting point, measure how the density of fog changes as the temperature of water in the bottle is varied. What atmospheric parameter might the temperature of the water in the bottle correspond to?

2. In Activity 4.3.2, you used a small hand pump to pump air into the flask before releasing it. Measure how the density of the fog changes as you vary the amount you pump. What atmospheric parameter might this correspond to.

3. Measure how the fog density changes as you vary the amount and type of condensation nuclei. Meteorologists have noticed that fog is often more dense in heavily populated areas. Do your results give any indication of why this might be?

UNIT D

BUOYANCY, PRESSURE, AND FLIGHT

DETAILED CONTENTS

UNIT D

BUOYANCY, PRESSURE, AND FLIGHT

Courtesy NASA

"The wind
Sweeps the broad forest in its summer prime,
As when some master-hand exulting sweeps
The keys of some great organ."

—*William Cullen Bryant*

---

0   OBJECTIVES

1. To develop an experimentally testable model of floating and sinking based on forces.

2. To compare the apparent weight of an object in and out of water and understand what determines the difference between the two.

3. To understand buoyancy and use it to explain why some objects sink while others float

4. To explore the difference between force and pressure.

5. To observe how gases respond to external forces.

6. To use the concepts of pressure and equilibrium to build a machine that can be used to lift very heavy weights with very little effort.

7. To explore how pressure differs at various locations in air and water

8. To investigate the role pressure differences play in buoyancy and in enabling birds and airplanes to fly

8. To learn more about the nature and causes of motion and the process of scientific research by undertaking an independent investigation.

0.1   OVERVIEW

The discovery that large hollow objects float dates back thousands of years. The ark is prominently mentioned in the Bible and Sumerian myths, and it is one of the most important creations in the history of civilization. Just imagine the possibilities opened up by the first boats! Suddenly people could gather their possessions and travel in groups across the water without fear of sinking. As boat-building techniques became more advanced, ships became larger and larger. Today, cruise ships are mini-cities, with movie theaters, ice-skating rinks, and restaurants capable of serving thousands of people at a time. It is one of the truly amazing features of nature that these gigantic boats can safely float on top of the water.

Most people have a fair amount of everyday experience with objects that float. Nevertheless, most people remain confused about even the basic elements of buoyancy. In particular, few people understand the role forces and *pressure* (another confusing concept) play in determining whether objects float or sink. Their role is quite subtle, and is often misunderstood. In this unit, we will make observations that help you build a fairly complete understanding of pressure and buoyancy.

A characteristic of the scientific process is the continuous refinement of initial, naïve intuitions. For example, you could probably rattle off a definition of "floating" without much difficulty, and in fact you will do this in the first activity. But this initial definition is unlikely to withstand careful scrutiny, and your definition will need to be refined as you look more closely at the phenomenon of floating. This is where you will begin your investigation—by finding inconsistencies (note these are not *mistakes*) and then refining your definitions.

Since the term pressure is often confused with the term force, we will attempt to develop a clear scientific definition of the term force. Then you'll measure an object's weight, both in and out of water, and try to understand why the weight appears to be different in water. You'll explore what role this phenomenon plays in determining whether an object sinks or floats. From there, you'll examine whether gases can exert a force on an object, which leads to a new property called pressure. Finally, we explore this property called pressure in the air and in the water to determine how barometers work and why airplanes fly.

©Andrew G. Wood/Photo Researchers

Section 1: How is force measured?   **Section 2:** What determines whether objects float or sink?   **Section 3:** Can a gas exert force?   **Section 4:** Why can airplanes fly?

Figure D-1: The main questions we will consider in this unit.

A word of advice: since the terms pressure, force, and buoyancy are probably terms you have heard and used many times, you undoubtedly have some understanding of what these terms mean, however vague. As we progress through the activities in this unit, we will be building scientific definitions that may or may not agree with your previous definitions. It is therefore very important to understand these scientific definitions. Since many of the concepts we will use rely on these definitions, a misunderstood definition can certainly impair your ability to understand a particular concept.

1   *FLOATING, SINKING, AND FORCES*

We begin this unit by considering what it means to float or sink. You are probably familiar with the everyday usage of the words *float* and *sink*, but how well do these definitions stand up to experimental testing? After observing how different objects behave when placed in or on a liquid we will try to develop a description based on forces that may explain all of your observations.

You may need some of the following equipment for the activities in this section:

- Large container filled with water (to immerse objects in) [1.1]
- Clay [1.1]
- Dixie cup [1.1]
- Wood block [1.1]
- Aluminum foil cut into flat squares [1.1]
- Small pieces of string (approx. 2 feet long) [1.2, 1.3]
- Springs (high spring constant, all identical) [1.2, 1.3]
- Meter sticks [1.2, 1.3]
- Rulers [1.2]
- Spring scales (10 N) [1.3]
- Block/mass with hook [1.3]
- Cart-track system [1.3]

1.1   "FLOATERS" AND "SINKERS"

We all know what it means to float, and that floating and sinking is not as straightforward as it first seems. After all, a large, ocean liner floats while a small marble ball sinks. One of the hallmarks of physics is the use of experimental observations to refine simple definitions. In this activity you will formulate a simple definition of floating and sinking, and the examine it more and more carefully.

Figure D-2: The Titanic was once the largest luxury liner in existence. It seems miraculous that such a heavy object, made mostly of steel, actually floats. At the same time if a young child throws a marble overboard from the passenger deck it will sink. (Hulton-Deutsch Collection/Corbis Images)

Activity 1.1.1 Defining *Floaters* and *Sinkers*

a) Have a quick discussion with your group (2-3 minutes) with your group about how you might describe an object that floats (a *floater)*. After your discussion, complete the following sentence: "An object that floats is one that…"

b) Now have another quick discussion (2-3 minutes) with your group about how you might describe an object that sinks, (a sinker). After your discussion, complete the following sentence: "An object that sinks is one that…"

c) Write your definitions on a separate piece of paper and exchange them with another group. Write the other groups' definitions for sinkers and floaters below.

d) Using the materials at your lab station, find as many exceptions to the other group's definitions as you can. Exceptions can include (but are no means limited to): objects that, by their definition, both sink and float and objects that seem to be defined as a *sinker* or *floater* but in fact do the opposite. List your exceptions below, write why they are exceptions, and be prepared to defend your exceptions in class.

e) Write new definitions for *floater* and *sinker* that are consistent with all of your observations, exceptions included.

As you have seen, often our first intuitive definitions, while agreeing with the majority of our general observations, produce contradictions when carefully tested. The process of science is frequently one of refinement; initial ideas are gradually modified and polished until they agree with more and more observations. As we proceed through this process of refinement it is often advantageous to focus on a particular aspect of the phenomenon we are investigating. At this point, we will narrow our investigation to focus on a single aspect of floating and sinking. That is, the pushes and pulls exerted on floating and sinking objects.

Competing Pushes or Pulls on Floating Objects

You have already observed what happens when various objects are immersed in a liquid. Some objects that are released begin to move up or down. Others remain at rest when released. Now we want to focus our attention on pushes and pulls that are acting on an object and how these might relate to whether the object floats or sinks.

Activity 1.1.2 Pushes and Pulls on Immersed Objects

a) Pick an object that you classify as a sinker. Hold the object at the top of the container and release it. Draw a picture of the object *immediately after release* and draw arrows to represent any pushes or pulls you think may be acting on the object. Label each arrow to identify the source of the push or pull it represents. Use the length of the arrows to indicate the relative strengths of the pushes and pulls. **Note:** Larger pushes/pulls should have longer arrows and smaller pushes/pulls should have shorter arrows.

Figure D-3: A rubber duck is a floater while a marble is a sinker. Why?

b) Now put a floater at the top of the water and release it. Draw a picture indicating any and all pushes and pulls acting on the object.

c) Repeat the above for a sinker and floater released from the *bottom* of the container.

d) The sinker at the bottom and the floater at the top are similar in that they don't move once released. How does your model for pushes and pulls link these two different scenarios? How does your model link the floater (sinker) released from the bottom (top) of the container?

1.2 QUANTIFYING PUSHES AND PULLS: FORCE

Our everyday experience tells us that if a resting object begins to move, it is either being pushed or pulled. Looking at sinkers and floaters in light of this everyday experience, we can say that sinkers are being pulled down while floaters are being pushed up. In order to continue our study of buoyancy, pressure and flight, we need to take an interlude to examine pushing and pulling in more detail.

The scientific term to describe a push or a pull is *force*. Everyday experience tells us that it takes a force to get a resting object to move. What happens when more than one force acts on the object at the same time? Imagine a tug of war. If we apply forces in opposite directions to an object at rest, in which direction will it move? What does your answer tell you about the nature of competing forces pushing or pulling on floaters and sinkers?

In this section we will focus on defining force more carefully and investigate ways of quantifying and measuring forces. This will provide you with the tools needed to investigate how competing forces might be used to explain floating and sinking behavior.

Unfortunately, the term "force" has many different meanings in the English language. This word also has a very specific scientific meaning, so we must be careful to understand precisely what it means before using it. This first activity will give you a chance to try and figure out what that meaning is.

Activity 1.2.1 A First Look at Force

a) Your group will be given a piece of string and some objects. This piece of string can be used to pull on an object. Your task is to determine a way of quantifying your pull. That is, determine a method of measuring a specified amount of pull by assigning some kind of strength value to it. Briefly explain your method below.

b) Test your method! Get a member of another group to come and follow the method you wrote above. You can't give the person any verbal help; your written instructions should suffice. If your instructions are insufficient, refine them in the space below.

If you found the above activity a little strange and confusing, don't feel bad. Although you can use a piece of string to pull on an object, it may not be the best tool to use to try and quantify or measure a specified amount of pull. One problem with using a piece of string is that once the string is taught, you can pull harder and harder and there is no noticeable change in the string until it breaks. Therefore, you might be tempted to use your sense of "feel" to try and determine how hard you are pulling. The problem is, everyone feels things differently and it is almost impossible to reliably pull with the same "strength" in different situations. Thus, we need to somehow figure out a way of assigning a "value" to the strength of the pull that is independent of our human senses.

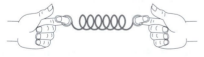

Figure D-4: What happens to the pull of the spring on your fingers as you spread your hands further apart?

Activity 1.2.2 Defining Force

a) Your group will be given several identical springs. Like the piece of string, these springs can be attached to something so that you can push or pull on them. Begin by placing one end of a spring around one of your fingers and have your partner place the other end around one of their fingers. Now gently move your fingers further apart and closer together. Does the pull on your fingers feel any different as you move your fingers further apart? Explain.

b) If your eyes were closed, how could you tell whether or not your fingers were close together or far apart?

c) Explain how you can use your spring to exert a pull on an object that is not changing. We call this a constant force.

d) Now, using a spring, work with your group to *define* a specific amount of pull (call it whatever you want). Write your definition below. That is, finish the following sentence, "We define one _____ to be ...." **Note:** Be specific. Uttering something like "We define one Godzilla to be 3 inches" is ambiguous.

e) Now explain how you could pull on something with two, five, or three hundred times as much pull as you defined in part d). Will any problems arise as you increase the amount of pull?

┌───┐
│   **Checkpoint Discussion: Before proceeding,**   │
│   **discuss your ideas with your instructor!**   │
└───┘

Developing a Standard Force Unit

As the last activity just demonstrated, it is not difficult to apply a constant pull using a spring. All you need to do is maintain a fixed length. Notice, however, that exactly what you decide to call "one unit of pull" may be quite different from what somebody else decides to call one unit of pull. This unit is completely arbitrary, in the same way that defining a length called an inch, a foot, a meter, or a furlong, is completely arbitrary. The important thing is that everyone agrees to use the same units or at least to be able to convert from one unit to another. This is what's known as creating a standard unit. As simple as this seems, developing a standard unit is an essential prerequisite to making quantitative observations.

The scientific term for a push or pull is force and there are a few different standard units that are widely used. In the United States, the most common unit is the pound. In the rest of the world, it is the Newton. The pound and the Newton are both units of force (with one pound equal to 4.45 Newtons). Thus, we can pull (or push) on something with a force of one pound or a force of one Newton.

Figure D-5: One Newton is roughly the force needed to hold up a small apple.

Representing Force with Arrows: Force Diagrams

When analyzing the forces on an object, it is often helpful to draw a picture of the object with arrows representing each of the forces. This is called a *force diagram*. In this type of drawing it is important to include all the forces acting directly on the object of interest. For example, if you were interested in the forces acting on the block of shown below, you would include the forces exerted on the block by the springs, but not the forces exerted by the hands, since the hands do not apply a push or pull directly on the block.

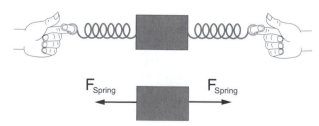

Figure D-6: Representing the forces on a block using arrows. The direction of each arrow indicates the direction of the force it represents, while the length of the arrow represents the strength of the force.

The direction of each force should be represented by the direction of its corresponding arrow and the relative strength of the force should be indicated by the length of the arrow. Thus, there are three rules for drawing the arrows in force diagrams.

1. Draw an arrow for each of the forces acting directly on the object in question, but do not draw arrows for forces acting indirectly on an object.

2. Draw each arrow so that it points in the direction of the force it represents.

3. Draw each arrow so that its length represents its strength relative to the other forces in the diagram.

1.3   HOW FORCES AFFECT RESTING OBJECTS

It seems obvious that an object that is sitting still will not spontaneously start moving. It takes a push or a pull. Consider what happened when you released a floater at the bottom of a container of water. Initially it wasn't moving but when you released the floater, it started moving upward. When it reached the surface of the water, it stopped moving. There is a difference between the situation at the top and bottom of the water. Understanding this difference will require further investigation.

Figure D-7: What happens to a stationary object when there are multiple forces acting on it?

In the activities that follow you will investigate the effect forces have on stationary objects. The conclusions you reach will provide you with a framework for analyzing floating and sinking. Your instructor may demonstrate these situations for you.

Activity 1.3.1 Forces on Objects at Rest

a) Suppose a cart is held at rest between two horizontal *balanced forces*—forces that have *equal strength* but act in opposite directions, as shown in the sketch below. On the lower sketch, draw arrows representing the strengths and directions of the forces acting on the cart. If nothing is changed, will the cart move? If so, in what direction

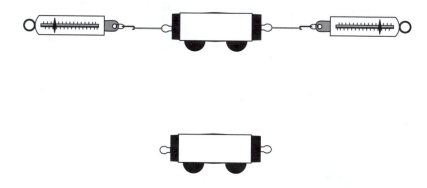

b) Now imagine that we cut the string that is applying the force on the right end of the cart, as shown in the sketch below. On the lower sketch, draw arrows representing the strengths and directions of the forces acting on the cart. Will the cart begin to move? If so, in what direction?

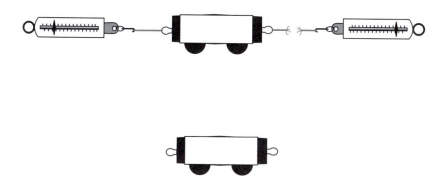

c) Now imagine that the cart is held at rest between two horizontal forces in opposite directions that have unequal strength. As in parts a) and b), draw the arrows that correspond to the forces acting on the cart immediately after the cart is released. Will the cart begin to move? If so, in what direction?

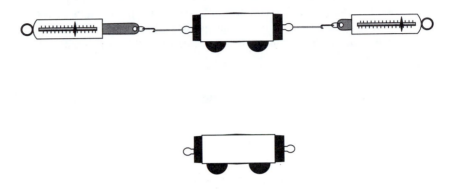

d) Now, try these experiments or watch your instructor demonstrate them. Describe your observations.

Balanced and Unbalanced Forces

In the previous activity we saw that in some situations an object can remain stationary when more than one force acts on it. One situation in which this occurs is when two forces of equal strength are acting in opposite directions. In this situation, we say that the forces are *balanced*, in other words, the total (or net) force acting on the object is zero. We also observed situations in which there were two forces of unequal strength acting in opposite direction. In this situation, we say the forces are *unbalanced*.

What happens when one of the two forces is stronger than the other? Your instructor will demonstrate a few situations that involve unbalanced forces.

Activity 1.3.2 Balanced and Unbalanced Forces

a) Consider the following statement by a student:

"If an object that is initially at rest remains motionless, then I can deduce that there must be no forces acting on the object."

Rebut this student's statement by setting up a situation in which a stationary object remains motionless despite having obvious forces applied to it. Use the objects and spring scales at your workstation to do this. Then rewrite this statement so that it is consistent with your observations.

b) Consider the following statement by a student:

"If an object that is initially at rest begins to move to the right, then I can deduce that there is a single force pushing (or pulling) the object to the right."

Rebut this student's statement by setting up a situation in which a stationary object begins moving to the right despite having an obvious force pushing or pulling it to the left. Rewrite this statement so that it is consistent with your observations.

As the previous activity suggests, the real power in understanding how balanced and unbalanced forces affect an initially stationary object is that we can infer whether or not an object is subjected to balanced or unbalanced forces simply by watching whether it begins moving or remains stationary. In fact, if we know one of the forces acting on an object, we can often infer the strength and direction of the other forces acting on the object. This is a technique we will use over and over throughout this unit.

Weight as a Force

Weight is another term (like float, sink, and force) that most people are quite familiar with. In fact, some of us are always worrying about our weight! It turns out that there are a number of ways to define weight. By now you should realize the importance of defining things carefully. In defining weight, we need to take into account that an object has weight because of the gravitational force exerted on it by the Earth. This force acts to pull objects toward the center of the Earth. If we were on a different planet that had a different gravitational force an object would have a different weight. Therefore, we define *weight* as the *gravitational force* acting on an object. For our purposes, we are only interested in situations on the surface of the Earth, where the gravitational force on a particular object is always the same and pulls the object straight down.

Activity 1.3.3 Inferred Forces

a)   Suppose a block of wood is held suspended at rest as shown in the sketch below. Are the forces on this block balanced or unbalanced? Give a reason for your answer.

b)   One of the forces acting on this block is the gravitational force, or the weight of the object. Draw arrows representing the object's weight on the diagram to below. Explain why there must be at least one other force acting on this object? What can you infer about the size and direction of this force?   Explain. **Hint:** Is anything touching the object?

c)   What do you predict will happen to the block immediately after the
     string is cut? After the string is cut, will the forces acting on the
     block be balanced or unbalanced? Draw arrows to represent the
     strengths and directions of the forces acting on the block after the
     string is cut. **Note:** The block is sketched by itself below to make
     it easier for you to draw your arrows.

d)   What do you think this activity has to do with sinking or floating?

Relating Balanced and Unbalanced Forces to Sinking and Floating

The power of the observations just made is not what it tells us about forces. The true
value is that it allows us to deduce the existence of new forces simply by observing the
motion of objects. If we observe an object that remains motionless, we can conclude that
there is no net force (either no forces are acting or the forces are balanced). Thus, if we
know the strength of the force in one direction, we immediately know the strength of the
force in the opposite direction. Conversely, if we see a stationary object begin moving
then we know the forces are unbalanced, and that the larger force acts in the direction the
object moves. We will now use this idea to investigate the forces that water exerts on
floating and sinking objects.

Activity 1.3.4 Forces on Floaters

a) Suppose a floater is released from the bottom of a cup of water, but has not yet reached the surface. We know that the gravitational force is acting on this object. We also know that water is touching the object so it is reasonable to infer that the water exerts a force on the object. What can you deduce about the direction and size (relative to the gravitational force) of this water force? Explain your answer. **Hint:** Are the forces acting on the object balanced or unbalanced.

b) Now, suppose that the object released in part a) is now floating. What can you deduce about the size of the water force now? Explain.

c) Since the gravitational force on an object near the surface of the Earth is always the same, what can you conclude about the size of the water force when the object is completely submerged compared to when the object is floating? Explain briefly.

d) Discuss possible reasons for any changes in the water force with
    your group, and list them below.

Checkpoint Discussion: Before proceeding, discuss your ideas with your instructor!

Now that we have a reasonable understanding of forces and of how balanced and unbalanced forces affect an object that is at rest, we are ready to apply these ideas to try and understand floaters and sinkers. In scientific terms, the *water force* you explored in the last activity is called the *buoyant force*. We begin in the next section by taking a quantitative look at this force and how it can be used to explain floating and sinking.

2	*BUOYANCY*

We begin this section by learning to use an electronic force sensor. This will allow us to measure the actual amount of force that is being exerted on an object. By comparing the force needed to lift an object when it is in air compared to when it is in water, we will be able to deduce the amount of force, if any, exerted by the water on the object.

You may need some of the following equipment for the activities in this section:

- MBL system [2.1-2.3]
- Force sensor (electronic) [2.1-2.3]
- Hanging mass [2.1]
- String [2.1]
- Tall tank of water for submerging objects [2.2]
- Small weights or metal shot (bee-bees) [2.2]
- Two small & one large water-tight plastic vials (approx. 70 ml & 200 ml). Fill each of the vials with shot to adjust its weight so that one of the small vials weighs approximately 1 N while the other small vial and the large vial each weighs the same amount of about 2.5 N. [2.2, 2.3]
- Overflow can and catch bucket [2.3]
- 500 ml beakers [2.3]
- A collection of small submersible objects of different sizes [2.3]

2.1   WEIGHING OBJECTS

The electronic force sensor you will be using works on the same principle as the stretched springs. When you pull or push on the sensor hook, an element inside the sensor is stretched or compressed. An electrical signal that depends on the amount of stretch or compression is sent through an interface box to a computer. The data collection software can then calculate the strength of the force applied to the sensor and display a real-time graph of how the force changes in time.

As you will see, force sensor readings depend on the spatial orientation of the sensor. The following activity is designed to help you understand this dependence and learn to account for it.

Activity 2.1.1 Playing With Force Sensors

a) Begin by opening the data collection software and without attaching anything to the force sensor, move the sensor hook so that it points up, then sideways, then down. Explain what you observe and how orientation changes might lead to uncertainties in your measurements.

Figure D-8: An electronic force sensor hook is placed in three different orientations. Even through there seems to be no force applied to the hook, the readings are different why? (Courtesy of Vernier Software and Technology)

b) Again open the MBL software and change the orientation of the force sensor, but this time either press the tare button on the side of the sensor (if one exists) or zero the sensor using a software command. Explain what you observe and how this would help you make a more accurate measurement.

c) Now try pulling and pushing on the sensor to see how it behaves. Briefly describe your observations including, for example, how you know whether you are pushing or pulling on the force sensor hook. **Note:** The sensor will only measure up to a specified amount of force. So if you push or pull too hard, the force readings will simply get stuck at a particular value even though the strength of the force keeps increasing.

d) Now attach a spring to the sensor and stretch the spring so that it pulls on the sensor with one of the force units you defined earlier. Remember to make sure that the sensor reads zero when it is in the proper orientation before attaching any force to it! Record how many Newtons your force unit is equivalent to. **Note:** One way to get a good force reading is to attach the stretched spring to the force and hold it very steady for about 5-10 seconds. This makes it much easier to read from the graph.

Using a Force Sensor to Measure Weight

In Activity 1.3.3, we learned that if an object is lifted and held steadily that the upward lifting force will exactly balance the downward force of gravity. Thus, we can measure the weight of an object simply by measuring the force necessary to lift it and hold it steady. The electronic force sensors you used in the last activity are ideally suited for this task. In the next activity, you will determine the weight of an object by hanging it from a force sensor and reading the force measured by the sensor.

Activity 2.1.2 Measuring the Weight of an Object

a) Measure the weight of an object using the force sensor. To do this, hang the object from the force sensor hook and hold the force sensor steady for about 5 seconds. Then record the force indicated on the computer screen. Do this three separate times and record your results below (don't forget units!). Are all the measurements exactly the same? If not, does this mean the weight of the object changes or does this tell you something about the force sensor? Briefly explain your answer. **Note:** Make sure to zero the force sensor when it is in the proper orientation before making your measurements.

b) Based on our conclusions from Activity 1.3.3, we know that the net force on our object is zero (make certain you understand this). Draw a force diagram of the object you weighed in part a). Carefully label all of the forces acting on the object with arrows.

Figure D-9: An object attached to a piece of string hangs from an electronic force sensor.

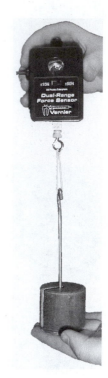

c) Now, place your hand under the object and push upward very gently as shown in Figure D-10. Record the force needed to lift the object as measured by the force sensor. Is the net force on this object still zero? Explain how you know. Also, draw a force diagram below, carefully representing the relative magnitude and direction of all three forces on the object.

d) We know that the gravitational force on the object does not change when we push up gently on it. In other words, the object's *true weight* does not change. Nevertheless, the *apparent weight* of the object—the force measured by the force sensor—has decreased. Explain how you can deduce the upward force of the hand on the object by measuring the *true* and *apparent weights* of the object.

Figure D-10: What happens to the apparent weight of an object when you push up on it gently?

The previous activity is loaded with information that you should talk over with your group and be ready to discuss with the entire class. Some questions you should consider are: (1) How accurate is the force sensor and how much can you trust a particular measurement? (2) How can you tell when the net force is zero and what information does this give you? (3) How can an observed change in one force indicate that another force acting on the same object has also changed?

> **Checkpoint Discussion: Before proceeding, discuss your ideas with your instructor!**

Measurement Accuracy

A point that is worth emphasizing is our inability to measure things with perfect accuracy. Thus, even if we know that an object weighs *exactly* 10 Newtons (10 N), we might measure it to be either 9.9 N or 10.2 N. The accuracy of our measurement will be limited by a number of factors including the type of measuring device and our ability to use the device and read it properly. It is impossible to eliminate these measurement uncertainties completely, so we must be careful when comparing two measurements. For example, the measurements of 9.9 N and 10.2 N are not the same, even though they were obtained by measuring the same object with the same force sensor. One unlikely conclusion is that the weight of the object is actually changing. Another is that there is a certain amount of uncertainty associated with the measurement process that prevents us from making a completely accurate measurement. So how can we make a more accurate measurement? One way is by making the measurement many times and taking an average of the results. This average will usually be more accurate.

In this unit, we will sometimes compare measurements made on two different objects that are expected to have the same weight. For example, suppose we measured a weight of 9.9 N for one object and 10.2 N for the other. Do these two objects have the same weight? Well, the answer depends. If you are using a measurement device that is accurate to within 0.01 N, then you can say that these objects have different weights. If, however, your measurement device is only accurate to 0.2N, then you cannot say with certainty whether these objects have the same weight or not. When the two measurements lie "within the uncertainty of the measurement device," we will often just assume they are equal (even though they may be different by a very small amount). It is important to keep this in mind as you make your measurements.

2.2   OBJECTS IN WATER

In the last activity you saw how an object's apparent weight changes when you pushed up on it. It takes less force to hold the object up. Similarly, it takes less force to lift objects when they are in water. Does that mean that they weigh less when they are in water? Well, no. The gravitational force on the objects does not change, thus their true weight remains unchanged. Still, the force needed to lift them while in water—their *apparent weight*— does change. In the next few activities you will investigate the weight of objects immersed in water.

Activity 2.2.1 A Sinking Object Under Water

a)   Imagine the following situation. An object that we know is a sinker is hung from a force sensor and slowly lowered into a tank of water. Predict what will happen to the force reading (increase, decrease, remain the same) as the object is being lowered into the water but is only partially submerged.

b)   Predict what will happen to the force reading (increase, decrease, remain the same) after the object is fully submerged and is lowered further into the water but does not touch the bottom.

c) Now try the experiment. Turn on the force sensor, zero it in the correct orientation, hang an object from it, and then *slowly* and *carefully* lower the object into a beaker of water. Continue lowering the object slowly even after it is fully submerged, being careful not to let it hit the bottom or sides of the beaker. Print out the graph or sketch it below. Label the graph to clearly indicate which portions of the graph correspond to when the object was completely out of the water, partially submerged, and fully submerged. Were your predictions in part b) correct?

d) Draw three force diagrams below, one corresponding to when the object is not yet in the water, one for when it is partially submerged in the water, and one for when it is fully submerged. Make sure your force arrows have appropriate lengths, indicating the relative magnitude of the forces.

e) Explain what you think is responsible for this decrease in the apparent weight of the object. Do you think there are also forces acting on the sides of the object? If so, why doesn't the object move sideways?

f) Why do you think the *apparent weight* of the object stops decreasing once the object is completely submerged?

Comments About the Forces on Sinking Objects

The fact that objects have a smaller apparent weight in water than out of water may or may not surprise you. For example, people notice that when they are swimming their bodies feel almost weightless. We call this weight-loss effect *buoyancy*. Now, it is possible that the Earth's gravitational force on an object is weaker in water and that is why the object loses weight. But since the weight of an object does not change when it's near water or when it's inside a submarine submerged in the water, this means the gravitational force would only be reduced when an object is physically in contact with water. This seems like a kind of arbitrary rule to associate with the gravitational force. Most people feel that since the water is actually in contact with the object, that it is more likely for the water to be pushing on the object in such a way as to reduce its weight. Now, since the water surrounds the object, the water pushes on the object at many places and in many different directions. However, the net effect of all this water pushing on the object is an upward force we call the *buoyant force*.

Activity 2.2.2 Measuring the Buoyant Force

a) Let's assume that the net effect of the water "pushing" on an object is to exert a force on it. Based on the measurements you made in the last Activity determine the *strength* of the buoyant force on the object when it is completely submerged. Explain your reasoning and show your work.

b) You have noted already that some objects float, others sink, and still others do neither. What characteristics of an object do you think will determine whether it will sink, float, or do neither? Are there any characteristics that will not matter? Discuss this with your partners. **Warning**: This is not a simple question! Consider a boat. It can float, but if you poke a small hole in the bottom, it can fill with water and sink.

Predicted Characteristics that affect buoyancy	Predicted Characteristics that don't affect buoyancy

---

Discovering Factors that Affect Buoyancy

In the last activity, you and your partners identified several characteristics of an object that might affect its buoyancy. In the next few activities, we will explore which factor actually affect an object's buoyancy. For example, it is possible that the color of the object has an effect on the buoyant force. Now, most people would probably agree that it seems very unlikely for color to have any effect on the buoyant force, and therefore, we would probably not waste our time carrying out an experiment to check. It is worth noting, however, that the only way to be *absolutely certain* that color is not a factor would be to carry out an experiment that tests this idea.

Since we only have a limited amount of time in this class, we will only test a few of the characteristics that are most commonly believed to affect the buoyant force. Since many people think that both the size (volume) and weight of an object may affect the buoyant force, you should test these factors. When testing the influence of different characteristics it is important to only change one factor at a time. In this way, *and only in this way*, can we be sure that any changes we observe are due to the characteristic under consideration.

Activity 2.2.3 Does Weight Affect Buoyancy?

a) Do you think that an object's weight will affect the amount of buoyant force it feels? Let's consider two objects that have exactly the same size and shape but not the same weight. Will they experience different buoyant forces. If so, will the heavier object or the lighter object have the larger buoyant force? Give a reason for your answer.

Figure D-11: These two objects have the same size, but one is heavier than the other. Will they experience the same buoyant force?

b) Now try the experiment. Your instructor will have some empty vials and small weights for you to use. Your objective is to determine if changing *only* an object's weight (without changing its size or shape) will change the buoyant force acting on the object. Discuss with your group how you will accomplish this task. Next, describe your experiment and record your results below. Try to keep your work organized so that you can refer back to it at a later time.

c) Make a short, concise statement supported with experimental evidence about whether an object's weight affects the buoyant force.

It is very important that you are confident with your results and fully understand what you have accomplished. A common mistake is to hurry through an experiment and simply write down the numbers without thinking much about what they mean. Now would be a good time to discuss with your instructor the procedure you used in the previous experiment, the results you obtained, and the conclusions you came to.

Checkpoint Discussion: Before proceeding, discuss your ideas with your instructor!

Our next task is to see whether the volume of the object affects the buoyant force. For this observation you will use two vials of different size and some small masses.

Activity 2.2.4 Is Volume a Factor

a) Do you think that an object's volume will affect the amount of buoyant force it feels? If so, will the larger object or smaller object have the larger buoyant force? Give a reason for your answer.

Figure D-12: These two objects have the same weight, but one is larger than the other. Will they experience the same buoyant force?

b) Now try the experiment. Your instructor will have some empty vials and small weights for you to use. Your objective is to determine whether changing *only* an object's size (without changing its weight) will change the buoyant force acting on the object. Discuss with your group how you will accomplish this task and then describe your experiment and record your results below. Try to keep your work organized so that you can refer back to it at a later time.

d) Make a short, concise statement supported with experimental evidence about whether or not an object's size affects the buoyant force.

Some Conclusions about Buoyancy

At this point, you should have reached some conclusions regarding buoyancy. First, it should be clear that the buoyant force always acts upwards on an object. We have also seen that the buoyant force increases as more and more of the object becomes submerged. But once the object is completely submerged, the buoyant force is constant and does not change no matter how deeply the object is submerged. Finally, the last two activities demonstrated that the weight of the object does not affect the buoyant force but the size (volume) of the object does.

2.3   HOW LARGE IS THE BUOYANT FORCE?

As scientists, we would like to understand everything there is to know about the buoyant force. So far, we have made some excellent progress, but we still can't predict how large the buoyant force will be on an arbitrary object. To understand this, we need to take a closer look at what's happening when an object becomes submerged. This is the topic of the following thought experiment.

Activity 2.3.1 An Interesting Observation

a)   The drawing below shows a container of water with nothing in it and then the same container after a small cube is placed in it. Notice that the water level is higher after the cube is placed in the water. This happens because the cube has to push some water out of the way to become submerged. What property of the object determines how much water is pushed above the original level? Be specific.

b)   If a different cube, having the exact same dimensions as the first but 10 times heavier (lead instead of wood) were dropped into a different cup containing the same amount of water as the first cup, how would the water level compare with that of the first cup?

c)   Here's an interesting coincidence! The amount of water that is pushed up (or *displaced*) by a submerged object depends on the object's volume but not its weight, *just like the buoyant force.* Perhaps the amount of water *displaced* is related to the buoyant force. This is an intriguing possibility. How might you connect the amount of water displaced and the buoyant force?

d)   Can you think of a physical reason why the water that is pushed up by the object might be responsible for the buoyant force?

If you don't understand how the immersion of an object results in some water being lifted above the original surface level, you should talk this over with your group and/or your instructor.

Now, since the object is responsible for displacing this water, it should seem at least plausible that this water might in some way be linked to the buoyant force. Furthermore, the lifted water will push down on the water below it with a force equal to its weight. Thus an attempt to link the displaced water with the buoyant force might start by examining the weight of the displaced water and comparing it with the buoyant force. Of course, at this point our explanation is only a conjecture. In order to either support or refute this conjecture, we need to perform an experiment.

The Overflow Can

In order to determine whether the force needed to lift the displaced (pushed aside) water is related to the buoyant force on the object doing the displacing, we need a way of collecting this water. A simple method for collecting water above a particular level is by using a cup with a hole in the side of it. If the water is filled up to the hole, then any additional water will drain out of the hole and can be collected. A schematic of this so-called *overflow can* is shown below with a smaller catch bucket used to collect any water that rises above the original water level.

Figure D-13: An overflow can and a smaller catch bucket can be used to measure the amount of water displaced by a submerged object.

We can use an overflow can to determine the weight of the water that is displaced by the submerged object as follows. Fill up the overflow can until water begins leaking out of the spigot (this is called *priming* the overflow can). After water stops leaking from the spigot, place an empty catch bucket under the spigot. Then, when an object is placed into the water in the overflow can, any water that's displaced by the object will be collected in the catch bucket. By using a force sensor, you can easily determine how much force is necessary to lift the displaced water and compare this to the buoyant force. Make sure you read the following instructions carefully and discuss with your group exactly how you will carry out the experiment. There is a lot going on!

Instructions:

1) Begin by priming the overflow can and then placing an empty catch bucket beneath the spigot. Attach a string to the catch bucket so you can lift it with the force sensor later.

2) The basic idea is to measure the weight of the object in air, then in water, then measure the weight of the displaced water. This can all be accomplished in one experiment, but you need to plan carefully.

3) Open your data collection software and set it to record force for at least 60 seconds.

4) After opening the MBL software, orient the force sensor as it will be when you make your measurements and zero it.

5) Start the force measuring software and hold the force sensor in the correct orientation with nothing attached to it. After about 5 seconds, lift the object with the force sensor and again hold it steady for about 5 seconds.

6) Next, lower the object into the water in the overflow can until it is completely submerged and hold it steady until no more water drains from the spigot. (You will need to be careful here to make sure that the object is completely submerged the entire time and that it is not resting on the bottom of the cup.)

7) Then, remove the object from the force sensor, hold the sensor steady for about five seconds and then lift the catch bucket full of displaced water and hold it steady for about five seconds.

8) Now you have all the information you need to determine both the buoyant force and the weight of the displaced water.

Activity 2.3.2 Weighing Displaced Water

a) Print out or sketch your graph of forces vs. time below and label
    each portion carefully. Record the buoyant force and the weight of
    the displaced water below and compare the results.

LARGER OBJECT: Force vs. Time

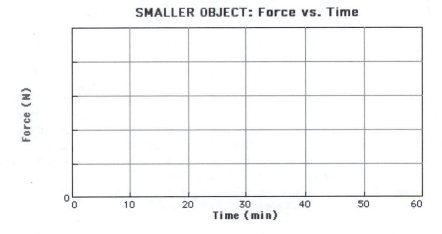

Buoyant Force (N) _____

Weight of Displaced Water (N) _____

b) Now repeat the entire experiment with a smaller object. Again,
    print out a copy for your activity guide and compare the buoyant
    force to the weight of the displaced water.

SMALLER OBJECT: Force vs. Time

Buoyant Force (N) _____

Weight of Displaced Water (N) _____

c)   Based on the results of these experiments, do you think there is any relationship between the buoyant force and the weight of the displaced water? Discuss this with your group and write down your idea as clearly as you can.

| **Checkpoint Discussion: Before proceeding, discuss your ideas with your instructor!** |

Archimedes Principle

The scientific law you have deduced in the previous activity is named after the Greek mathematician Archimedes (287-212 B.C.) and is called *Archimedes' principle*. Archimedes' principle states that the buoyant force acting on an object is equal to the weight of the liquid displaced by the object. Although we have only verified this for water, it turns out to be true for other liquids as well (this would make a nice project).

Warning: The amount of water that is displaced by an object may in fact depend on the orientation of the object when it is submerged. For example, a drinking cup, when fully submerged in an upside up orientation will fill up with water and therefore, it will not displace very much water. However, if this cup is first turned upside down and then fully submerged, then there will be air trapped inside the cup which will cause the cup to displace much more water than in the other orientation. This adds a subtle complication that we will not explore (another excellent project!). In this unit, we will only concern ourselves with "sealed" objects, which displace the same amount of water regardless of their orientation.

THE LEGEND OF ARCHIMEDES AND THE CROWN

Many of the principles of buoyancy you have been studying were discovered by

©Archivo Iconografico, S.A./Corbis Images

Archimedes – a Greek mathematician. One of the legends about his exploits involves buoyancy. According to the legend, King Hiero commissioned a goldsmith to make him a crown. Heiro provided the goldsmith with a particular weight of pure gold to make the crown. Upon receipt of the crown, Heiro suspected that the goldsmith had replaced some of the gold with the same weight of a less valuable material like silver. The king asked Archimedes to determine whether or not the crown was made of pure gold. While sitting in a bath Archimedes discovered that the amount of water that spilled over the sides of the bath had about the same volume as the parts of his body that were submerged. He was so excited about this realization that he ran naked through the streets of Syracuse yelling "Eureka, Eureka" (I found it, I found it). According to the legend, Archimedes used his discovery along with a procedure similar to the one in Activity 2.3.2 to expose the goldsmith's fraud. The legend suggests that Archimedes submerged an amount of gold equal to the weight of the crown into a container of water that was completely full. He then submerged the crown into the water and discovered that it displaced more water than the pure gold. As a result, Archimedes was able to detect the fraud.

Archimedes' principle is very useful because it allows us to determine, among other things, whether or not an object will float or sink.

Activity 2.3.3 Will it Sink or Float?

a) If an object is completely submerged in a liquid, there are two forces acting on the object, the downward gravitational force from the Earth and the upward buoyant force from the water (which equals the weight of the displaced water). Draw a force diagram for this object and explain how these forces determine whether this object will float or sink.

b) What would happen to the object if the gravitational force was exactly equal to the buoyant force (this is called being neutrally buoyant)? Explain briefly (and be careful).

As the previous activity demonstrated, determining whether an object floats or sinks is a relatively simple manner (remember, we are only discussing *sealed* objects here). If the object's weight and the buoyant force are not the same, then the forces are unbalanced. Recall that when we have unbalanced forces, an object released from rest will begin moving in the direction of the larger force. Thus, if the object's weight is larger than the buoyant force, it will begin moving down (sink), and if the buoyant force is larger than the object's weight, it will begin moving up (float).

But how do you know what the buoyant force will be? You could go through the overflow can experiment, but is there an easier way? The answer to this question is yes, and it makes use of Archimedes' principle. This is the topic of the following activity.

Activity 2.3.4 Calculating the Buoyant Force

a) Since the buoyant force is equal to the weight of the displaced liquid, you can calculate the buoyant force by determining the weight of the displaced liquid. Now, if the object has a fairly regular shape, you can compute its volume by making certain length measurements (for example, length, width, and height if it has a box-like shape). Explain how this volume can be used to determine the amount of displaced water.

b) Different objects will, in general, have different volumes which will result in different amounts of displaced water. For example, one object might displace 15 cm^3 of water while another displaces 72 cm^3 of water and yet another displaces 158 cm^3 of water. Thus, to determine the buoyant force that acts on these three objects, it appears as though you need to measure the weight of 15 cm^3 of water, then 72 cm^3 of water, and then 158 cm^3 of water. But that seems like an awful lot of work. Together with your group, devise a method whereby you only need to measure the *weight* of a *specific* amount of water *once* and that will then allow you to calculate all three buoyant forces (or any others for that matter). Explain your method below.

c) Now carry out this process by measuring out a specific amount of water and measuring its weight. Record your results below. **Hint:** It is much easier and more accurate to measure the weight of a large amount of water (100 cm^3 – 200 cm^3) and then multiply or divide your result to get the weight of the specific amount you are after. **Note:** 1 ml = 1 cc = 1 cm^3.

d) Now use your result to calculate the buoyant force on the three objects given in part b) assuming they are completely submerged. If each of these objects has a weight of one Newton, comment on whether they will float or sink.

e) For an object that is actually floating, we know that the forces will be balanced. This means the buoyant force will be equal to the weight of the object. But we just observed that when a floater is completely submerged, the buoyant force acting on it is larger than its weight. Explain how the buoyant force on a floater goes from being greater than its weight to being equal to its weight as the floater goes from being totally submerged to floating on the surface (only partially submerged).

f) For any floaters from part d), calculate the exact volume of the object that is submerged when the object is floating.

```
╔═══════════════════════════════════════════════════════════╗
║      Checkpoint Discussion: Before proceeding, discuss      ║
║                your ideas with your instructor!             ║
╚═══════════════════════════════════════════════════════════╝
```

Specific Weight of a Liquid

In the previous activity, we needed to calculate the weight of three different volumes of water. Rather than actually weighing those particular volumes of water, we found that we could just as easily weigh one specific volume of water and then multiply by the correct conversion factor to get the weights we were interested in. This is a situation that arises quite frequently in the sciences. In this situation, we could measure the weight of say, 1 cm^3 of water. This is something that only needs to be measured once. Then, if we want to know the weight of 7 cm^3 of water, we merely multiply the weight of 1 cm^3 of water by 7. And if we need to know the weight of 54.8 cm^3 of water, we multiply the weight of 1 cm^3 of water by 54.8. Because the weight of 1 cm^3 of a substance is so useful, it is given a special name. It is called the *specific weight* (meaning the weight of a specific amount) of the substance and is usually denoted by the symbol γ. Using the specific weight, we can write the buoyant force of an object more simply as $F_B = \gamma V$, where V is the volume of the liquid displaced by the object.

The specific weight of water is given by 0.0098 N/cm^3 (or, equivalently, 9,800 N/m^3). This means that each cubic centimeter of water has a weight of 0.0098 Newtons (or that each cubic meter of water has a weight of 9,800 Newtons). Notice that the specific weight is a property of a substance (water, mercury, lead, etc.) and not of an object. Thus, we might say that an ice cube (an object) has a weight of 0.25 N, but ice (the substance) has a specific weight of 8,987 N/m^3.

Mass and Density

In this unit, we have focused our attention on the concept of force and found that we could easily measure the weight of an object with a force sensor. A concept that is closely related to an object's weight is its *mass*. Most people are familiar with the term mass, but it is actually a rather abstract concept. Mass is measured in kilograms and can be defined as an object's resistance to acceleration or as a measure of the amount of matter present in an object. Near the surface of the Earth, there is a simple relationship between an object's weight and its mass given by $W = mg$, where $g = 9.8$ m/s^2 is a constant called the acceleration due to gravity. Because weight is the quantity that we actually measure, there is no reason to introduce the more abstract concept of mass.

In analogy with the specific weight of a substance, one can define the "specific mass" of a material, which is more commonly called the *density* of a material. Density tells you the mass of a specific amount of material. Like specific weight, density is a property of a material and not a property of an object. Thus, we might say an ice cube (an object) has a mass of 25.5 grams, but ice (the substance) has a density of 917 kg/m^3. For the same reasons as with the term mass, we will not use the term density in this Activity Guide but will instead focus on the more meaningful term specific weight.

Important Note! A common mistake that many students make is to say something along the lines of "if an object is more dense than water it will sink and if it is less dense than water it will float." First off, this statement is technically incorrect. As described above, the term density refers to a property of the material the object is made from and not the object itself. But beyond that, this statement is simply not true. Take steel, a material that is obviously more dense that water. It is quite easy to make an object made entirely out of steel float. All you have to do is form the steel into the shape of a large, thin, "salad bowl." Then, the weight of the water displaced by the object will be larger than the weight of the object itself. This is precisely the condition under which an object will float.

Activity 2.3.5 The Titanic: Floater and then Sinker

a) Explain in your own words why it is possible for the Titanic to float while a marble that is thrown overboard sinks.

b)   In 1912 the Titanic hit an iceberg, ruptured its hull, and sank. Over 1500 passengers were killed in this tragic accident. Using what you know about buoyancy, explain why the ship sank when its hull was ruptured.

_____

3	**PRESSURE**

In the last section, we explored the concept of force and applied it to try and understand why some objects sink while others float. We begin this section by considering how forces feel. We will see that the concept of force is very useful in some instances but not so useful in others. We will define a new quantity called *pressure* that is related to force, but more useful in certain situations. In fact, that the concept of pressure can be used to deepen our understanding of buoyancy. Additionally, the concepts of both pressure and buoyancy are very useful in learning about how airplanes and birds can fly.

You may need some of the following equipment for the activities in this section:

- Bungee cords or rubber tubes: ~2' long [3.1]
- Small (< 1/4") & large diameter (> 1/2") metal rods or wooden dowels [3.1]
- Ruler with cm scale [3.1]
- Glass syringes [3.1, 3.2]
- MBL system [3.3, 3.4]
- Force sensors [3.2]
- Small weights: 100 g & 200 g [3.2]
- Plastic tubing: ~15 cm long [3.2]
- Magdeburg hemispheres [3.3 Demo]
- Plastic syringes: disposable [3.3, 3.4]
- Pressure sensors [3.3]
- Plastic tubing: ~1 meter long [3.4]
- Tall (at least 50 cm) tanks of water with different diameters [3.4]

3.1   ARE EQUAL FORCES ALWAYS EQUAL?

If you have ever walked around in your bare feet, you will have undoubtedly noticed that it is not uncomfortable to walk on grass, but it can be very uncomfortable to walk on a surface that has gravel or loose rocks all over it. Why exactly it this? After all, your feet are supporting your body's weight in both situations. The following activity explores this idea by applying the same force to our bodies in slightly different ways and observing the difference.

The basic idea is to attach a bungee cord (i.e., a big spring) to a metal rod and pull the rod against your body with a constant force. Then we'll repeat the experiment using a different rod but pulling with the same constant force. (Alternatively, you might be able to pull on both rods simultaneously.) This is a little difficult to accomplish by yourself, so you might need to work in groups of two. Also, it is best to have the rod pushing against a part of your body that is not very fleshy. Your shins work quite well.

Figure D-14: A shin "pain" meter: A wooden dowel is held in place across your shin by a bungee cord wrapped around the leg of a chair. How does changing the dowel size affect what you feel?

Activity 3.1.1 Force & Pressure: What's the Difference?

a) Attach a bungee cord to a large-diameter metal rod (the larger the better, but at least a half inch) and pull with this rod against your shins with a fixed amount of force. Then change to a small diameter rod (the smaller the better, but no more than a quarter inch) and pull with the same fixed force. Explain your observations below. (If you don't think you feel any difference, try pulling both rods against your shins simultaneously using two bungee cords.)

b) Are the two rods pushing against your shin with the same amount of force? Explain exactly how you know this. Is there a difference in the way your shins feel in each case? Discuss these questions with your group and explain your ideas below.

c) If you were going to try and make these rods "feel" the same against your shins, would you pull harder or smaller on the smaller rod?

    d)   Consider the following discussion between two students:

> **Student A:** *"Of course the small rod is pushing with more force. It hurts more! How could it feel different if it was pushing with the same force?"*

> **Student B:** *"The two rods are pushing with the same force since the bungee cords are stretched the same amount each time. Where would the extra force come from?"*

What would you say to these students? If you agree with Student A, how do you answer the question posed by Student B about the origin of the extra force? If you agree with Student B, how do you answer the question posed by Student A about the extra push?

This activity illustrates just how different the effect of the *same* force can be. Even though the same force is applied in both situations, the result can feel dramatically different. As you may have guessed, it is the amount of *physical contact* with your body that distinguishes just how "painful" a given force might be. A more striking example is to consider how it would feel if you were to lean against a flat wall compared to leaning against a wall that has a nail sticking out of it. In this case, the difference in pain would be quite substantial! Clearly, our bodies are sensing something different than force. Precisely how this comes about is the subject of the next section in which we look for a quantitative connection between how a force is applied and the effect it can have. We will accomplish this by examining the forces exerted by a gas. Before doing so, it will be useful to give a very simple description of three common forms of matter.

Solids, Liquids, and Gases

Although we won't go into too much detail about the differences between solids, liquids, and gases, a few of the more obvious features are worth pointing out. Basically, a solid is a material that maintains its volume and shape. A liquid is a material that maintains its volume yet conforms to the shape of a container that it's placed in. A gas is a material that expands to fill the volume of the container that it is placed in, thereby changing both its volume and its shape. Of course, there are other materials that don't really fall into these three simple categories. Jell-O, for example, maintains its volume but can change its shape fairly readily. Sand can be poured like a liquid from a bucket and formed into a solid castle. We won't concern ourselves too much with these interesting materials and will instead focus our attention on gases and liquids.

Figure D-15: The eruption of Old Faithful at Yellowstone Park in Wyoming involves three different forms of matter. Hot water (a liquid) from the core of the Earth is trapped under rocks (a solid). As the water escapes from a fissure in the rocks the pressure on it is suddenly reduced and it vaporizes into steam (a gas). The steam encounters cooler air (another gas) and condenses into water droplets (liquid) which we see as a cloud. (©Toyohiro Yamada/FPG International)

One important feature of gases, liquids, and solids follows directly from our simple definitions above. As you will see, it is relatively easy to change the volume of a gas by compressing it or allowing it to expand. However, solids and liquids maintain their volume and it is almost impossible to compress or expand them by any significant amount. We will explore this property of liquids a little bit later.

Most people do not find the behavior of liquids to be all that different from ordinary objects. After all, when you pour a liquid out of a cup, it falls to the floor like most objects. But the behavior of gases is a little different. If you pour a gas out of a container, it doesn't fall to the floor like a solid or a liquid. In some cases (for example with helium) gases almost seems to defy gravity. Of course, there's no reason to think that gravity doesn't act on the gas the same way it acts on other objects. So why doesn't a gas fall to the ground like other objects? Well, there must be some other force that can affect a gas but has little effect on a liquid or a solid. We will return to this interesting point later in the unit.

Forces from Gases

Next, we consider forces exerted not by springs, strings, or water, but by the air itself. For many of the activities in this section, you will be using glass syringes. Glass syringes are used because there is very little friction between the piston and the rest of the syringe. This means that we can neglect friction and focus our attention on what else is going on. Unfortunately, precisely because there is very little friction, *you must be very careful when handling glass syringes. It is easy to have the piston slide out and fall onto the floor and break. So please, be careful!*

Activity 3.1.2 Gas Force I

a) Begin by taking a glass syringe and pulling the plunger (which we'll refer to as the piston) about half way out. Now place your finger over the opening and describe what happens when you gently push or pull on the piston. Try to describe your observation in terms of forces and directions. That is, when you push or pull on the piston, explain whether or not there are any forces acting on the piston, and if so, in what directions do they act?

b) Now try to explain what you think is actually exerting the force on the piston and whether it is physically *pushing* or *pulling* (or both). Does your answer depend on whether you are pushing the piston in or pulling the piston out? Explain briefly.

Figure D-16: What happens when you hold your finger over the opening of a syringe and try to push or pull on the piston?

c) If you believe the air pushes on the piston, explain how you think this pushing occurs. (For example, is there any contact between the air and the piston or is the air capable of pushing without contact?) Also, if you believe the air pulls on the piston, explain how you think this pulling occurs. (For example, is the air "sticky" like glue or does it have "hooks" that attach to the piston?)

d) Now remove your finger and push the piston all the way in so that there is no air inside the syringe. (If you're bothered by the small amount of air in the tip of the syringe, your instructor can help you remove that air with a small hand pump.) After placing your finger over the opening, try pulling on the piston once again. Describe your observation and explain whether you think there is something *pushing* or *pulling* on the piston. Specifically, if something inside the syringe is pulling on the piston, what could it be? If there is nothing inside pulling on the piston, why does it bounce back when you release it?

e) Does this cause you to re-consider your answer to part b? Explain briefly.

```
┌─────────────────────────────────────────────────────────┐
│                                                           │
│   Checkpoint Discussion: Before proceeding, discuss       │
│           your ideas with your instructor!                │
│                                                           │
└─────────────────────────────────────────────────────────┘
```

Most students initially suspect that the gas inside the syringe is somehow responsible for both pushing and pulling on the piston. This is not unreasonable. However, by performing the same experiment with *no air* inside the syringe, we have demonstrated that, at least in this situation, the air inside the syringe cannot be responsible for pulling on the piston (there is no air in the syringe!). Thus, a pulling force is not necessary to cause the observed behavior of the piston. This suggests that there must be something *pushing* on the piston and the only material that is in contact with the outside of the piston (besides our fingers) is the air in the room.

Now, although the air inside the syringe may have different properties than the air in the room, they are both the same substance and should therefore behave in a similar manner. Thus, if the room air is pushing on the piston, then it seems reasonable to expect the air inside the syringe is also pushing on the piston (although possibly with a different strength). In fact, maybe all gases are only capable of pushing (but not pulling) on objects. It's still possible that gases are capable of both pushing and pulling, but it would be much simpler if gases could only push. Let us see if we can explain our earlier observations using this simple idea. In the previous activity, we observed that when there is *no air* inside the syringe and your finger is placed over the end, the piston feels a force pushing it into the syringe. We will assume that the air in the room is responsible for pushing the piston into the syringe.

Activity 3.1.3 Can Pushing Result in Pulling?

a)   Now image holding the syringe horizontally and pulling the piston about halfway out before placing your finger over the tip of the syringe (see sketch below)

If you then release the piston, what will happen? Since the room air hasn't changed, there must be a force pushing the piston into the syringe. An arrow representing this force has been drawn with the diagram of the piston below. By applying the concept of balanced and unbalanced forces, determine the size and direction of the force that the air inside the syringe must be exerting on the piston and draw an arrow representing this force on the diagram below. Explain your answer.

Room Air

b)   Now imagine that you have pushed the piston in while still holding your finger over the end of the syringe. (See sketch below)

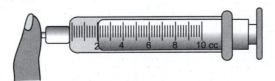

On the diagram below, draw an arrow representing the force you exert by pushing on the piston. By applying the concepts of balanced and unbalanced forces, determine whether the size of the force exerted by the air inside the syringe has changed from what it was in part a). Draw an arrow representing this force on the diagram below and explain your reasoning. **Hint:** Since there is no change to the air in the room, the room air force pushing the piston into the syringe will not have changed.

Room Air

c) Now imagine that you have pulled the piston out while still holding your finger over the end of the syringe.

On the diagram below, draw an arrow representing the force you exert by pulling on the piston. By applying the concepts of balanced and unbalanced forces, determine whether the size of the force exerted on the air inside the syringe has changed from what it was in part a). Draw an arrow representing this force on the diagram below and explain your reasoning. **Hint:** As before, since there is no change to the air in the room, the room air force pushing the piston into the syringe will not have changed.

Room Air

d) Based on your arguments above, determine a simple rule that describes how the amount of force exerted by a fixed amount of air can become larger or smaller depending on the volume of the vessel it is contained in. Describe your rule below.

e) Discuss with your group why the confined gas has more space might behave differently than when the gas has less space. Explain your ideas below.

Pushing and Simplicity

This activity demonstrates that it is indeed possible to explain our earlier observations without claiming that a gas can pull on an object. While it is conceivable that a gas could exert a pull through some yet unexplained manner, it is far simpler to limit gases to pushes. Here it is not difficult to imagine how a gas might push on an object it is in contact with (say, by bumping up against it). In general, when faced with two competing ideas, both of which can correctly explain the experimental observations, we are guided by a principle known as *Occam's Razor*. The principle of Occam's Razor states that when all else is equal, we should choose the simpler of two descriptions. In this situation, we opt for the simpler idea that a gas is only capable of pushing on an object.

Notice that to explain our observations in the previous activity, we were led to the conclusion that a fixed amount of gas will exert more (pushing) force when confined to a smaller volume and less (pushing) force when confined to a larger volume. This is an important idea that can be used to explain the following interesting phenomenon.

Figure D-17: Using a simple billiard ball model of a gas, it is easy to imagine how a gas might exert a pushing force. On the other hand, this simple model doesn't suggest any straightforward way that a gas could exert a pulling force.

Activity 3.1.4 What Holds the Piston in Place?

a) Hold a glass syringe upright and pull the piston about halfway out and release it (without placing your finger over the tip). Notice that the piston slides back down into the syringe. Now repeat the procedure but this time, place your finger over the end of the syringe before releasing the piston. Notice that the piston will no longer slide back down into the syringe. Explain from this observation which force is larger, the force from air outside the room acting down on the piston or the force from air trapped inside the syringe acting up on the piston. If it helps, draw a force diagram of the piston (don't forget to include the weight of the piston).

b) Recall that when the syringe is being held sideways, the two air forces have the same size (since the piston won't move when released). Based on your answer to the last question and the results of the previous activity, has the volume inside the syringe changed from just before you released the piston? Explain your answer.

c) Now turn the syringe upside down while still holding your finger over the end. Again, the piston does not slide out of the syringe. Explain from this observation how you can tell which force is larger, the force from air outside the room acting up on the piston or the force from air trapped inside the syringe acting down on the piston. If it helps, draw a force diagram of the piston (don't forget to include the weight of the piston).

d) Based on your answer to the last question and the previous activity, has the volume inside the syringe changed at all? Explain your answer. **Hint:** The force from the room air has not changed.

Inference

The point of the previous activity was to yet again point out *inferences* that you could make about the physical situation (volume of gas) from a careful consideration of what the relative forces must be. If the piston isn't moving when released from rest then there must be no *net* force acting on it. Thus, when the syringe is held upright with your finger over the end, the volume of the gas must go down just a little bit in order to counterbalance the force from the room air and the weight of the piston. While this change in volume is not obvious, (we could if we looked very closely), we know that it *must* have occurred. You now have the ability to explain why you are able to lift liquid by holding your finger over the end of a straw. This is a useful question to discuss with your group to test your understanding of the material.

Checkpoint Discussion: Before proceeding, discuss your ideas with your instructor!

3.2 THE SYRINGE MACHINE

As we discovered in Activity 3.1.1, the same force can feel very different depending on how the force is applied. In the following activity, we will see an interesting application of this phenomenon, which in turn will lead to the definition of a new quantity.

Activity 3.2.1 The Syringe Machine

a) Imagine two syringes, one of which is larger than the other, connected together by an airtight tube as shown in Figure D-18. We call this contraption the *syringe machine*. If you push down on one of the pistons, the other one will be forced to move up. Will the piston move the same distance or will one piston move further than the other? How is this related to the size of the syringes?

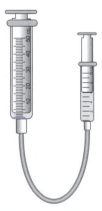

Figure D-18: The Syringe Machine.

b) Now imagine putting a small weight of say, 1 Newton (a mass of about 100 grams), on the piston of the smaller syringe. Because the two syringes are connected, this would cause the larger piston to move upwards. But suppose you were going to use your finger to push down on the *large* piston to hold up the small weight. How much force do you think you would need to apply with your finger? **Note:** A qualitative answer is fine.

c) How hard would you have to push if the 1 N weight was placed on the large piston and you were pushing down on the *smaller* piston? Try to give a reason for your answer.

d) Now set up the experiment using a 100 or 200 gram mass. First, place the weight on the large piston and lift it three or four times by pushing down on the small piston. Then place the weight on the small piston and lift it three or four times by pushing down on the large piston. Does the weight feel heavier in one situation and lighter in the other? Record your observations below. Do you think this effect is related to the relative distances the pistons move? Explain briefly.

e) Now, it should be clear that the actual weight of the object (and therefore the downward force exerted by that object) is the same in both situations. Discuss with your group why you think the force you need to exert to lift the weight changes depending on which piston the weight sits on. Can you predict how your applied force would change if you made one of the syringes longer or shorter? What about making the syringe wider or thinner? Explain what logic you are using to make your predictions as best you can. **Hint:** How does the gas exert a force on the pistons?

Clearly, there is something very interesting going on here. As a scientist, we would like to understand this curious behavior so that we can predict what will happen in different situations. Therefore, we need to make some quantitative measurements of this situation. But first, let's take a closer look at the syringe machine.

How Force Gets Through the Gas

When you push down on one of the pistons, the other piston is pushed up. The downward force on one piston somehow gets through the gas and results in an upward force on the other piston. Of course, the first piston doesn't actually push on the second piston, it only pushes on the gas that it is in contact with it. So one simple way of imagining this process is as follows. The first piston pushes on the gas next to it, and this gas pushes on the gas next to it, and this process continues all the way through the tube until the gas is finally pushing up against the second piston. This analysis leads one to the conclusion that the important part of the syringe machine is the area of contact between the pistons and the gas! **Note:** This is what we call the *cross sectional area* of the piston. This should remind you of Activity 3.1.1 where two different sized metal bars were pulled with equal forces against your shins.

Quantitative Investigation of the Syringe Machine

It's now time to search for a quantitative understanding of the syringe machine. To do so, we will compare the forces applied to the two pistons in the syringe machine and compare them with the cross sectional areas of the pistons.

Activity 3.2.2 The Syringe Machine Revisited

a) Let us begin by making a prediction. Imagine taking a known weight of 1 Newton and placing it on the larger piston in a syringe machine. Assuming the smaller piston had a cross sectional area that was half the size of the larger piston, predict what force would be necessary to lift the weight by pushing on the smaller piston. Explain the reasoning behind your prediction.

Small Cross Section

Large Cross Section

b) Carefully measure your two pistons and determine how much smaller (or larger) one is than the other. You may want to use a "caliper" to measure the diameter and then recall that the area of a circle is equal to πr^2, where r is the radius of the circle. Show all your work!

Figure D-19: Pistons with small and large cross sectional areas.

c) Now measure the weight of the object to be placed on the larger piston in the syringe machine and record your result below. You can do this by hanging it from a properly zeroed force sensor. Also, write down your prediction on how much force will be necessary to lift this weight using the syringe machine.

d) Finally, place the weight on the larger syringe and use the force sensor to lift the object on the syringe machine. It is best to push down on the piston until the weight has been lifted and then hold it steady for a few seconds to get a good reading. Repeat the measurement four or five times to make sure you get a reproducible result and record your force below. Does your measurement agree with your prediction? Explain briefly.

e) What do you think you would measure if you placed the weight on the smaller piston and lifted it by pushing down on the larger piston? Make a prediction and then test it by performing the experiment. Caution: it is *much* more difficult to get a reproducible reading when pushing down on the larger piston. You need to be very careful when performing the experiment, and even still, the results are sometimes ambiguous.

> ## Checkpoint Discussion: Before proceeding, discuss
> ## your ideas with your instructor!

The preceding activity demonstrated that the strength of an applied force can be modified using a suitably configured syringe machine. In this configuration, a force is applied to a piston that, in turn, pushes on a confined gas, that then, pushes on another piston. The resulting force on the second piston is not necessarily of the same magnitude or in the same direction as the force applied to the first piston. One might ask, "How does the air know how to do this?" The simplest answer to this question is that the air *doesn't* know how to do this. In fact, the air doesn't know anything and simply has the same properties throughout the syringe machine. After all, if you were a little piece of air, how would you know if you were in the tube near the larger piston or near the smaller piston?

A Property of a Gas

If we assume that the air in the syringe machine has the same properties throughout, then the two forces on either side of the syringe machine must have the same effect on the gas. Thus, although the forces being applied on the two sides of the syringe machine are not the same, when applied through different cross-sectional areas, they have the same effect on the gas. In some sense, they "feel" the same to the gas.

Activity 3.2.3 Force and Area

a) Recall that it took much less force to lift the weight when you pushed on the smaller piston than when you tried to lift the same weight by pushing on the larger piston. In some sense, we could say that a smaller area requires a smaller force to have the same effect. Is this behavior similar to or different from the experiment where you pulled metal rods against your shins? Explain briefly. **Hint:** You might want to review Activity 3.1.1, particularly question c).

b) In the last activity, we demonstrated that if one piston was three times larger (in cross sectional area) than the other, then it would take three times as much force on the larger piston to balance the force on the small piston. This implies that if you were to increase the force *and* the cross-sectional area of the piston by the same amount, then the effect on the gas would be the same. Working with your group, try to determine a way of combining force and area mathematically so that the result does not change if you multiply both the force and the area by the same (arbitrary) factor. This means that if you double or triple both the force and the area, then your mathematical combination should have the same value as it had initially. Show your work. **Hint:** You might want to try adding, subtracting, multiplying, or dividing.

c) Next, using your force and area measurements from Activity 3.2.2 part d), calculate the mathematical combination you just postulated and show that the result is the same for each side of the syringe machine (at least, within experimental uncertainties). Show the steps of your calculation.

> ## Checkpoint Discussion: Before proceeding, discuss your ideas with your instructor!

Defining Pressure

In the previous activity, there are many ways of combining force and area so that the result is not changed when both quantities are multiplied by an arbitrary number. For example, the quantity F/A is one possibility, as is A/F or $\sqrt{A/F}$ or many other combinations. In fact, there are an infinite number of ways of combining force and area such that the result is not changed when both quantities are multiplied by an arbitrary number. So which is the correct one? The answer is that there is no "correct" one! We are *creating* a definition for a property of the gas and we can define it to be anything we want. Of course, there is no reason to make an overly complicated definition, so we may as well choose the simplest combination that works. This leaves us with two choices, F/A and A/F. To choose between these two combinations, we go back to our experience with the metal rods against our shins and recall that the same force felt more painful when the area of contact was smaller. Thus, if we want our definition to correspond in some way with that experience, we should choose the combination that gets larger when the area gets smaller and the force is held fixed. The leads to a property we call *pressure*, defined as

$$P = \frac{F}{A}$$

3.3   MEASURING PRESSURE

In the above discussion of pressure, we made an important change in focus. Previously we had only concerned ourselves with the force that a gas could exert on a surface. The only way to measure this was to measure the "counter-force" needed to balance this force. Now, however, we are discussing quantities involving the gas itself. The gas can produce a force on any surface in that it is in contact with. While pressure is an indication of how strong that force will be, it is important to notice that pressure is not the same thing as force. Force, for example, always acts in a particular direction, while pressure has no direction associated with it whatsoever. It is the air that will exert a force on an object as a result of some contact between the air and the surface of the object. Thus, unless the gas is physically in contact with something, it cannot exert a force on it. The physical property of a gas that describes how it will exert a force on a surface is what we call pressure.

It is worth mentioning two features about the relationship between force and pressure. The first follows directly from the definition of pressure. If the pressure of a gas is the same everywhere, then the force exerted by this gas on an object depends only on the area of contact between the gas and the object. This is one reason why pressure is such a useful quantity, and it makes the operation of the syringe machine easier to understand. *If the pressure of the gas is the same everywhere in the syringe machine, the gas will exert a larger force on the piston with the larger cross-sectional area.* This is exactly what we have observed.

The second feature has to do with the direction of the force. Remember, pressure does not have a direction, but the force exerted by a gas does. We have already determined that a gas can only push on an object. It turns out that a gas can only push "straight into" an object. That means the force exerted by a gas is always directed straight into, or perpendicular to, the surface that the gas is in contact with.

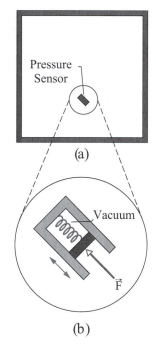

(a)

(b)

Definitions

How exactly do we measure pressure? Basically, you measure the amount of force exerted on a given area of surface and then you divide the force by the area. It's really that simple! There is a slight problem, however. You need to make sure that there is no force being exerted on one side of your measuring device. The way this is normally accomplished is to take a small syringe-like object and evacuate all the air from inside of it. Once the air has been evacuated, there will be no force being exerted from the inside of the syringe. To measure the force exerted from the outside, a small spring is placed inside the syringe. The amount the spring gets compressed is then related to the total amount of force being exerted by the gas. This is shown schematically in Figure D-20.

Figure D-20: A schematic representation of a pressure sensor. (a) in a gas or liquid (b) magnified. The measured pressure will be the same no matter what direction the sensor points in.

You also need to be aware of the many different units of pressure that are used throughout the world. One common unit of pressure is called the Pascal and is defined as one Newton per square meter. That is, a gas has a pressure of one Pascal if it exerts a force of one Newton when in contact with an area of one square meter. This is the unit we will be using in class. But just like the units of force, there are many other units for pressure. You might hear pressure readings given in units as *pounds per square inch*, *atmospheres*, or *inches of mercury*. For example, measuring the pressure of the outside air on a typical day might lead to a pressure of 1 atmosphere, 14.55 pounds per square inch, 29.92 inches of mercury, or 101,000 Pascals. These are all approximately the same pressure reading.

Activity 3.3.1 An Electronic Pressure Sensor

a) Unlike force sensor, pressure sensors are not sensitive to their orientation and therefore do not need to be zeroed. Use a pressure sensor to measure the pressure of the air in the room at a number of different places (near the floor, near the ceiling, etc.). Record your observations below. Does the pressure appear to depend on position in the room?

b) From your measurements in part a), calculate the amount of force exerted on one square meter of a surface in the room. Calculate the amount of force being exerted on one square centimeter of a surface in the room. What is the approximate force exerted on the door in the room?

c) Calculate the force exerted by the room air on a single piece of paper. How are you able to lift the paper so easily if this force is pushing on the paper? (And no, it's not because you're really strong!) **Hint:** Hold a piece of paper vertically in the air. If the paper does not begin to move to the left or the right, there must not be any net force acting on the paper. How can this be?

The first thing you should notice about the pressure sensor is that, like the force sensor, the measured values "bounce" around a little. This demonstrates that the pressure sensor, like any measuring instrument, cannot give you a perfectly accurate measurement. Typically, an inexpensive pressure sensor is accurate to about 1%, or about 1 kPa. The next thing to notice is that the value of the pressure doesn't seem to change (at least not more than 1 kPa) depending on where you are in the room. More accurately, all we can really say is that any pressure variation in the room is less than the precision of the force sensor. This supports our assertion that the gas in the syringe machine has the same properties, and in particular, the same pressure, throughout.

The Force Exerted by the Air

The previous activity demonstrated that the air in the room is capable of exerting extremely large forces. In fact, a typical 8½"x 11" sheet of paper has over 1,000 pounds of force acting on it from the room air. While that may sound impressive, everyone knows you easy it is to lift a piece of paper. So why don't we notice this large force from the air? Well, as you probably figured out, the room air is in contact with both sides of the piece of paper and therefore exerts equal forces in opposite directions, Thus, those forces balance out and we don't notice them at all.

To observe these very large forces exerted by the air, we need to remove the air from one side of an object, so that there is an air force on one side, but not the other. Unfortunately, this is not an easy task. One reasonably simple way of demonstrating the air force is by using a popular apparatus called Magdeburg Hemispheres (See Figure D-21). These hemispheres are placed together and then a pump removes the air from inside the sphere. Since there is no air inside, there will be no force pushing out on the hemispheres from the inside. There is however, still air in the room that pushes on the outside of the hemispheres, holding them together with an enormous force. If the diameter of the sphere is only 10 centimeters, it would take approximately 500 pounds of force to separate the two hemispheres. If the diameter of the hemisphere were 1 meter, the force required to separate the hemispheres would be over 50,000 pounds. Your instructor may demonstrate this for you. It is truly impressive!

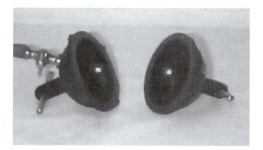

Figure D-21: Magdeburg Hemispheres are used to demonstrate the enormous strength of the force exerted by the air. (Courtesy Charles Johnson, North Carolina State University)

Activity 3.3.2 Pressure in the Syringe

a) Connect a syringe to the pressure sensor and try pushing and pulling on the piston of the syringe. Describe the pressure readings in relation to the pressure of the room when you push or pull on the piston.

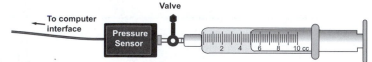

b) Do these results agree with the simple rule you developed back in Activity 3.1.3? Explain briefly.

c) Use the results of your pressure measurements in the room and inside the syringe to explain why the piston behaves the way it does when you push or pull on it.

d) Now take the syringe and without pushing or pulling on the piston, measure the pressure while holding it right-side up for a few seconds and then upside down for a few seconds. Repeat this a few times and describe your measurement. Do these measurements help justify your analysis in Activity 3.1.4? Explain briefly.

We noted back in Activity 3.1.3 that decreasing the volume a gas is allowed to occupy resulted in larger forces while increasing the volume resulted in smaller forces. As you just observed with a pressure sensor, we can also state that the pressure of a gas increases when the volume decreases and the pressure decreases when volume increases. Although we have only determined a qualitative relationship between pressure and volume, determining the exact relationship between pressure and volume for a gas would make a great project!

Notice that in the previous activity, the two ends of the piston are in contact with gases that have different pressures. This leads to unbalanced forces. Now, if the cross-sectional areas that are in contact with the two ends of the piston are equal, then this result is easy to understand. Since $F = PA$, if the areas are equal, then the larger pressure will result in a larger force. But at first glance, it doesn't look like the cross-sectional areas on the two ends of the piston are the same. It turns out, however, that the effective cross-sectional areas on opposite sides of an object are always the same, regardless of what the object looks like. Because of this fact, we can always determine the direction of air force acting on an object just by considering the pressures on opposite sides of the object. The following activity demonstrates a simple application of this idea.

Activity 3.3.3 Pressure and Wind

a) Imagine a stationary object in the shape of a cube sitting in a room where the air has the same pressure everywhere. Draw in the force arrows that act on this object and explain whether or not this object will remain stationary. Explain briefly (For simplicity, you only need to consider the left and right acting forces.)

b) Now consider the same stationary object in a room where the pressure on the left side is larger than the pressure on the right side. Again, sketch in the force arrows (left and right only) and explain whether this object will remain stationary or not.

c) Now imagine that the object in the previous question is nothing but a small parcel of the air itself. The movement of this parcel of air is what we would call wind. Now assume that you step outside and feel a wind blowing. What can you conclude about the air pressures based on your observation of this wind?

The previous activity demonstrates a very simple idea. When air has the same pressure on two different sides of an object, the force on each side of the object is the same and these forces are balanced. That means that the net force on the object is zero. If, on the other hand, the pressures are not the same on either side of an object, then the forces will be unbalanced and an object initially at rest will begin moving in the direction of the larger force. This means that a gas that has different pressures at different places will cause forces that move the air around until the pressures are all the same everywhere.

Gases and Equilibrium

An amazing feature of gases is that they naturally seek to equalize the pressure throughout. That means that if the pressure is *not* the same everywhere, then the gas will re-adjust itself until the pressure *is* the same everywhere. After a short time, the gas will have no macroscopic motion (although there may be motion on a microscopic level), which we refer to as a gas in *static equilibrium*. This leads to the following simple rule:

> When a gas is in static equilibrium, the pressure is the same everywhere throughout the gas.[14]

This simple rule is very useful and applies to a gas regardless of the shape of the container.

Keep in mind that the preceding discussion only applies to gases that are "connected" in some way. If the gases are not in contact, there is no way for them to exert forces on each other. Therefore, an open container will reach an equilibrium in which the gas inside the container is at the same pressure as the gas in the room. However, a gas inside a sealed container may or may not have the same pressure as the air in the room, even if they are both (separately) in equilibrium.

3.4   HYDROSTATIC PRESSURE

Our definition of pressure stemmed from our experiments with the syringe machine, which used air as a way of transferring force from one piston to another. Although this was a useful way of introducing the concept of pressure, it was defined in such a way that it would be useful in other contexts as well. For example, the term pressure is useful in describing how our bodies experience forces. Recall that equal forces do not necessarily

---

[14] Technically, this is only true if there are no external forces acting on the gas. When there are external forces acting, there can be pressure gradients that don't give rise to moving air. Gravity is an example of an external force, and in fact, the air on the surface of the Earth does not have the same pressure everywhere, even when there is no wind blowing. However, the pressure variation is very slight and is not measurable in the classroom. (Exploring this phenomenon would make a great project!)

feel the same to our bodies, and in fact, the pain we experience seems to be more closely related to the concept of pressure than to force. This suggests that the concept of pressure might be useful whenever we are dealing with forces and surfaces.

Our initial investigations in this unit revolved around the buoyant force that water exerts on objects. Since this force is being applied by the water onto the object, we can re-examine this topic from the perspective of pressure. To do so, we need to study the pressure of a liquid. As with a gas, we will only be examining the pressure when the liquid is in static equilibrium (i.e., when it is not moving). If the liquid is water, then it is said that we are studying *hydrostatic* pressure.

There are similarities and differences between gases and liquids. One of the defining characteristics of a gas is the fact that it is easily compressed or expanded. We observed this when we placed a finger over the end of a syringe and then pushed or pulled on the piston. A liquid, on the other hand, maintains its volume, which means it cannot be compressed or expanded. The following activity examines this briefly.

Activity 3.4.1 Compressing a Liquid

a)   Fill a plastic syringe that has a sealed end about half and half with water and air. *(Do not use a glass syringe for this!)* Then, push and pull on the piston. Now take some of the air out and repeat the experiment. Finally, remove all of the air and try once again. Record your observations below.

The Incompressibility of Liquids

You may be a little surprised that you could not move the piston when the syringe was filled only with water (if you *could*, then there was still some air inside, try it again). Unlike gases, liquids are extremely difficult to compress or expand. We call them *incompressible.*[15] This raises an interesting question. When a gas was compressed, we saw that its pressure increased. But if a liquid cannot be compressed, then you can't really change the amount of liquid that is in contact with a particular surface. Does that mean the pressure of a liquid will always be the same? To help answer this question, let us analyze the buoyant force acting on an object in a little more detail.

Figure D-22: A plastic syringe with its end sealed is filled with a mixture of air and water. How compressible are the contents with air present? With air absent?

Activity 3.4.2 Revisiting Buoyancy

a)   Assume an object in the shape of a cylinder (with height h and cross sectional area A) is held by a thread and completely submerged in a liquid. Let's assume that a liquid behaves just like a gas in that the pressure is the same throughout the liquid and the

---

[15] Of course, liquids aren't truly incompressible, but the amount of force required to compress them even the tiniest amount is extraordinarily large. For example, if you applied about 1,000 pounds of force to your syringe, the volume of the water inside would only be reduced by about 1%. Of course, the syringe would probably explode long before you could apply that much force to it.

force exerted by the liquid is directed straight into a surface. Using these assumptions, decide whether or not there would be any net force acting on this object? Explain your answer completely. **Hint:** Use the definition of pressure and deal with the up/down forces separately from the left/right forces.

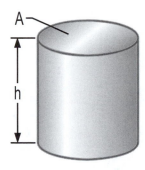

Figure D-23: An object that has the shape of a cylinder of height h and cross sectional area A.

b)   Knowing that the buoyant force (net force of the water on the object) acts in the upwards direction, what can we conclude about the size of the (water) forces acting on the top and bottom surface of this object? What can you conclude about the sideways forces? Explain briefly.

c)   You should have come up with contradictory answers to questions a) and b). This signifies that one of the initial assumptions must be incorrect. In this case, we find that the pressure *cannot* be the same everywhere throughout the liquid. Explain what you can conclude about the pressure in the liquid.

d)   What do you think would happen if you measured the pressure at different depths beneath the surface of the water? Discuss this with your group and make a prediction of the shape of the graph below. You may not be able to include actual numbers, but you should try to get the shape of the graph correct. **Hint**: The pressure at the surface of the water is equal to the pressure of room air. This is approximately one atmosphere, where

$$1.00 \text{ atm} = 1.01 \times 10^5 \text{ N} / \text{m}^2 = 14.7 \text{ lb/in}^2$$

Pressure vs. Depth

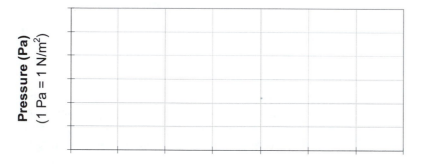

Depth Under the Surface of the Water (m)

Buoyancy and Hydrostatic Pressure

Notice that our knowledge of the buoyant force has allowed us to conclude that the pressure cannot be the same everywhere in a liquid. Having initially assumed that the pressure is the same throughout the liquid leads to the incorrect conclusion that there is no net force exerted by the water on the object. Since this contradicts our buoyant force experiments, we know that the assumption cannot be true. This kind of technique, where you assume one thing and then show that it leads to a contradiction (and therefore the assumption must be wrong) is used quite frequently in the sciences.

Measuring Pressure Under Water

In order to check your pressure prediction from the previous activity, we need to find a way of measuring pressure under water using our pressure sensors. **Caution:** *Our pressure sensors are designed to work only in air, so be very careful not to get water into the sensor!* Even though we cannot get water into the actual pressure sensor, we can attach a long tube to the pressure sensor and then, by placing the tube into water, we can measure the pressure at different depths.

Activity 3.4.3 Measuring Pressure Under Water—I

a)   Begin by filling a tall container (at least 50 cm in height) with water and attaching a long tube to the pressure sensor. First, measure the pressure of the room air and the pressure right at the surface of the water (this is accomplished by placing the end of the tube so that it just touches the water). Record your results and state any measurable differences.

b) Next, measure the pressure under the surface of the water by pushing the tube all the way into the water and then removing it. Repeat this a few times and describe your observations. Also, try moving the tube laterally (sideways) when it is under the surface and describe what you observe.

c) Notice that some water creeps into the tube when it is under water. Does this observation alone allow you to conclude anything about whether the pressure in the water was changing? Explain. **Hint:** What is happening to the gas in the tube?

Hydrostatic Pressure and Depth

You should have observed that the pressure increases under the surface of the water. In fact, the further below the surface you go, the larger the pressure becomes. You should also have observed that as long as you maintain the same depth below the surface, the pressure does not change. This is a very important observation that applies to all liquids as long as they are not physically separated from each other. In addition, your measurements showed that the pressure right at the surface of the water is the same as the pressure of the room air above the surface. This is a general feature that is true for all liquids.

Our next task is to make a series of measurements at specific depths under the surface of the water. In order to make accurate measurements, it is important to know precisely what depth we are measuring the pressure at. Shown below is a sketch of the pressure sensor tube under the surface of water. As you saw in the last activity, some water has crept into the bottom of the tube. This raises the question. Are we measuring the pressure at depth A or depth B? Let's find out!

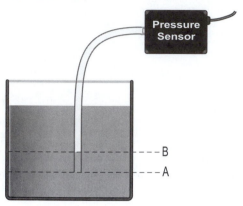

Figure D-24: This drawing depicts a clear, flexible tube that has been placed underwater to measure pressure. Is the pressure being measured at level A or level B?

Since we know that the pressure under water depends on depth, it should be clear that there is a difference in pressure between the liquid at level A (the bottom of the tube) and the liquid at level B (where the air meets the water inside the tube). Now, we also learned in the last activity that the pressure at the surface of a liquid is the same as the pressure of the gas above the liquid. Therefore, the pressure of the air inside the tube is the same as the pressure at the surface of the liquid inside the tube. This is the pressure at level B. But the pressure of the *gas* inside the tube (which is the same everywhere) is precisely the pressure that is measured by the pressure sensor. Thus we conclude that:

> *The pressure sensor measures the pressure of the liquid at level B, not at the bottom of the tube!*

The reasoning that leads to this conclusion uses much of what we've learned about pressure in gases and in liquids. It is worth thinking about this until you fully understand it. If you have questions, be sure to ask your instructor.

Activity 3.4.4 Measuring Pressure Under Water—II

a) Using a tall (as tall as possible) container of water to make a series of measurements of the pressure at different depths below the surface. Make sure you measure the correct depth below the surface. If you have any doubts, ask your instructor. Try to make at least five measurements including one right at the surface. (Make sure your distance measurements are reported in meters and your pressure measurements are reported in Pascals.)

Depth (m)	Pressure (N/m^2)

b) Next create a graph of your pressure vs. depth data. If you have graphing software available you may want to print out a graph of your data and affix it below. If your data seems to lie more or less along a line, draw a straight line that best fits your data.

Pressure vs. Depth

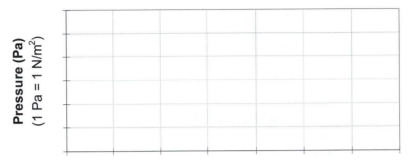

Depth Under the Surface of the Water (m)

c) Find an equation for the straight line that best fits your data above. Write your equation in terms of the physical variables P (*pressure*) and d (*depth*) as opposed to x and y. Try to give a physical interpretation for the slope of this line. **Hint:** Recall that the slope is the "rise over run." What do the "rise" and "run" represent on this graph?

d) Explain why this line does not pass through the origin and try to give a physical interpretation for the y-intercept. That is, what does it mean?

e) Use the equation you found above to calculate the depth for which the pressure would be twice as large as the pressure of the air in the room.

Checkpoint Discussion: Before proceeding, discuss your ideas with your instructor!

Connecting Buoyancy and Pressure

The equation you found for the pressure under the surface of water can be written in the following form, $P = P_{air} + \gamma d$, where P_{air} is the pressure of the air above the water, d is the depth below the surface of the water, and γ is the slope of the line, which we found in the previous activity. We can now completely explain the buoyant force in terms of this difference in pressure.

Let's derive and equation showing how the pressure differences on the top and bottom of a cylindrical object that is completely submerged under water can be used to determine the buoyant force on the object.[16] The height of the object is given by the difference in depth between the top and the bottom of the cylinder so that $h = d_{bottom} - d_{top}$. This is shown schematically in Figure D-25.

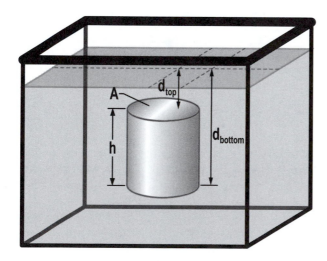

Figure D-25: A cylindrical object of height h and cross sectional area A is suspended in water. Its top is a distance d_{top} below the surface and its bottom is a distance d_{bottom} below the surface.

---

[16] Although we won't prove it here, the results of this analysis is valid for an object of any shape.

Activity 3.4.5 Pressure and Buoyancy

a) Write an expression for the force on the top surface of the cylinder due to the water. **Note:** Remember that the pressure under the surface of the water is given by the pressure of the air above the water P_{air} plus a constant γ times the dept under the surface d.

$$F_{top} = P_{top}A =$$

b) Write an expression for the force felt by the bottom surface of the cylinder due to the water.

$$F_{bottom} = P_{bottom}A =$$

c) The buoyant force is the *net* upward force on the cylinder due to the object. Using your answers to parts a) and b), write an expression for the buoyant force of the water in terms of the area A and the height h of the object. **Hint:** What is $d_{bottom} - d_{top}$?

$$F_B = F_{bottom} - F_{top} =$$

d) A cylinder's volume is equal to its height h multiplied by the area of its top and bottom faces, $V = hA$. Using this, re-write your expression for the total buoyant force so that it includes the cylinder's volume instead of its height and area.

e) Does the expression you wrote in part d) agree with your observations on buoyant force from Section 2 (especially Activity 2.3.2)? Explain any agreement or disagreement.

Connecting Pressure and Buoyancy

In the activity above, you derived an expression for the buoyant force that depended only on the object's volume. This equation says that the buoyant force is equal to the slope of our pressure-depth line multiplied by the volume of the object (or the volume of water displaced by the object). This equation should look somewhat familiar. It is exactly the same equation that we developed when we found that the buoyant force was equal to the weight of the liquid displaced (Activity 2.3.4). In that case, the constant was the specific weight of the liquid and here, it is the slope of our graph of pressure vs. depth. Because these two formulations for the buoyant force must be the same, this tells us that the slope of our line is simply the specific weight of the liquid!

We have now come full circle and connected our ideas about the buoyant force with our ideas about pressure. From this, we see that the buoyant force on an object is intimately related to the fact that the pressure changes with depth in a liquid. Without this pressure change, there would be no buoyant force.

Buoyancy in a Gas?

As we have seen in class, the pressure of a gas is the same throughout the room, at least within the accuracy of our pressure sensors. Since there is no pressure change for a gas, then the previous analysis says that a gas will not exert a buoyant force on an object. On the other hand, we also know that the buoyant force is equal to the weight of the liquid displaced. Since an object placed in air will certainly displace some air, it makes sense that there will be a buoyant force equal to the weight of air displaced. Although we will not pursue this question further, an investigation of this topic would make a great project!

4	THE REAL WORLD: BAROMETERS AND AIRPLANE FLIGHT

We are now in a position to apply what we've learned about forces and pressure to some real-world situations. In particular, we will look at the operation of a mercury barometer and the basic principles underlying airplane flight.

You may need some of the following equipment for the activities in this section:

- Graduated cylinders [4.1]
- Clear plastic tubing or clear straws [4.1, 4.2]
- Empty aluminum cans [4.2]
- One-holed rubber stoppers [4.2]
- 3"x 5" Index cards and thumbtacks [4.2]
- Air source [4.2]
- MBL system and Pressure Sensor [4.2]
- Plastic Tubing [4.2]
- Compressed air source [4.2]

4.1   BAROMETERS: MEASURING THE PRESSURE OF THE ATMOSPHERE

We have discussed a gas as a material that will expand to fill whatever container it is placed in. A natural question to ask is why the air on the Earth doesn't "expand" away from the Earth? After all, there is no container to hold it in. This is a good question! We have an atmosphere on Earth because gravity pulls the air towards the surface of the planet and prevents it from escaping. This pulling action is a little bit like someone pushing a big piston down from above. Thus, gravity acts to compress the air and increase its pressure. The result is a force on everything that the atmosphere comes into contact with.

The weather is intimately related to the atmosphere. Therefore, knowledge of atmospheric pressure is a key ingredient to understanding and predicting the weather. Today, measuring the atmospheric pressure is as simple as plugging in a pressure sensor. But how was pressure measured hundreds of years ago?

Pressure and Water Levels

There are a number of methods for measuring pressure. We will be discussing one that was designed to measure the pressure of the atmosphere—a mercury barometer. Before getting into the details of this device, we need to understand a few preliminaries.

Figure D-26: An old fashioned mercury barometer was used for over a century as a weather forecasting instrument. Modern barometers are now mostly electronic. (© O. Comitti & Son Ltd., London)

Activity 4.1.1 Lifting Liquid with a Straw

a) Fill a graduated cylinder or other clear cup with water. Take a clear plastic tube (or straw) and place it in the water. Without placing anything over the end of the tube, slowly pull it upwards without taking it completely out of the water. Describe the water levels inside the tube compared to outside. Is the pressure of the air inside the tube the same or different from the pressure of the air outside? Give a reason for your answer.

b) Now push the tube all the way down into the container. Place a finger over the end of the tube and then slowly lift it up without taking it completely out of the water. Describe the water level inside the tube compared to outside. Is the pressure of the air inside the tube the same or different from the pressure of the air outside? Give a reason for your answer. **Hint:** Recall Activity 3.1.4.

c) Now release your finger and then without moving the tube down, place your finger back over the top and then slowly lower the tube into the water. Describe the water level inside the tube compared to outside. Is the pressure of the air inside the tube the same or different from the pressure of the air outside? Give a reason for your answer. **Hint:** is the air being compressed or expanded?

d) Based on these simple experiments, do you think there is a relationship between the pressure of the air in the straw and the water level in the straw? Explain briefly.

_____

As you learned in the text following Activity 3.4.3, there is no relationship between the pressures in liquids that are in physically separate containers. But, if the liquid is in a single container, then we know that the pressures at the same level are all the same. Furthermore, since we know how the pressure under the surface of a liquid depends on depth ($P = P_{air} + \gamma d$), we can determine a relationship between the water levels and pressures. This is the underlying principle that governs how a barometer works.

A Simple Barometer

As you have probably known since you were a child, you can raise the liquid level in a straw by sucking a little air out of the straw and then quickly placing your finger over the end. The more air you remove, the higher the liquid rises. Would this process continue indefinitely? That is, if you had a tube that was extremely long, could you keep raising the level of the liquid in the tube by removing more and more of the air out of it? The answer to this question is no. Once *all* of the air has been removed from the tube, the liquid will not rise any higher. When all the air has been removed and the tube has been sealed, we have what is commonly called a barometer. A sketch of this is shown below.

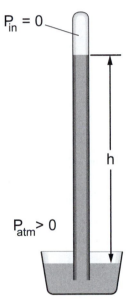

Figure D-27: A simple barometer can be used for measuring the pressure of the atmosphere. All of the air has been evacuated from the top of the tube. Since there is no gas to exert a force, the pressure in this section is zero.

Activity 4.1.2 How a Barometer Works

a)   If one uses a barometer as shown in Figure D-27, would it be reasonable to use the height of the liquid column as a measure of the pressure? That is, is there anything wrong with the statement "the pressure today is 15 feet of water?" Explain why or why not.

b) At the end of the last section, we experimentally determined an equation for the pressure of a liquid underneath the surface. This pressure is given by $P = P_{air} + \gamma d$. In this equation, P_{air} is the pressure of the air above the surface of the liquid, d is the depth below the surface, and γ is the specific weight of the liquid. This equation is valid at both of the liquid surfaces in our barometer. Let's apply it to the higher surface and use the height of the column, h, as the depth to find the pressure at point S. Begin by writing down the pressure of the air above the top surface and the dept of point S.

$$P_{air} =$$

$$d_S =$$

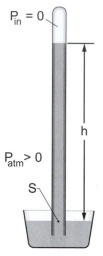

Figure D-28: A simple barometer.

c) Now substitute these into the equation for the pressure in a liquid to obtain the pressure at point S, in terms of the height of the column of liquid and the pressure at the surface of the liquid.

$$P_S =$$

d) Notice that point S is at the same level as the surface open to the atmosphere. What does this tell you about the pressure at point S and the pressure at the open surface? Explain briefly.

e) Finally, putting all these pieces together will give a very simple formula for the pressure of the atmosphere.

$$P_{atm} =$$

f) If the atmospheric pressure were to suddenly drop, would the column of liquid rise or fall? How is this reflected in your answer to part f)?

g) Now use the above formula to calculate the height of a column of water that would be supported by an atmosphere with a pressure of 100,000 Pascals. Approximately how many feet is this? Also, calculate the height of a column of mercury that would be supported by this same atmosphere. Approximately how many feet is this? Show all work. **Note:** The specific weight of water is 9,800 N/m^3 and the specific weight of mercury is 133,300 N/m^3.

h) Which would make a better choice if you were building a barometer, mercury or water? Explain briefly.

Checkpoint Discussion: Before proceeding, discuss your ideas with your instructor!

Although it may seem strange, there is nothing inherently wrong with quoting the pressure of the atmosphere in terms of the height of a column of liquid. In fact, a common unit of pressure is "inches of mercury" (in.-Hg.). The meaning of this should now be clear. It tells you the height of a column of mercury that can be supported by the atmosphere. As with any scientific definition, describing pressure this way is only useful if everyone agrees to use the same definition. Unfortunately, pressure is one of those quantities in which *many* different definitions are in widespread use. So don't be surprised if you hear pressure readings quoted in Pascals, millibars, pounds per square inch, torr, inches of mercury, or even atmospheres.

4.2   AIRPLANES, FLIGHT, AND THE PRESSURE OF WIND

We began this unit with a discussion of buoyancy, an understanding of which made possible major advances in human travel (e.g. the construction of ships). We conclude by investigating the relationship between pressure and another remarkable advance in the technology of travel, flight. Have you ever wondered how something as massive as a jumbo jet can stay in the air? Even though these airplanes weigh many tons, in flight, they are supported only by the air. In this section, we will explore how air can exert the forces necessary to keep an airplane in the air.

Activity 4.2.1 An Airplane Wing

a)   The figure below shows a side view of an airplane wing traveling through the air. In order for the airplane to stay in the air, in what direction must the air exert a force? Explain briefly.

b)   If the force exerted by the air is due to pressure differences, explain where (around the wing) the air pressure must be relatively large and where it must be relatively small to produce the forces needed to keep the plane in the air. Label areas of high pressure and areas of low pressure on the diagram below.

c) Can you think of any reason why the pressures would be different at different points around the airplane wing? Explain briefly.

_____

Here we have used a common observation (airplanes fly) in addition to our knowledge of forces and pressure to deduce that there *must* be different pressure on the top and bottom of an airplane wing. At this point, the reasons that these pressures should be different are still a bit of a puzzle. However, we do have one clue. Airplanes move very quickly through the air. So, perhaps the moving air around the wing is somehow responsible for this pressure difference. The following activity explores this idea by taking a look at the pressure exerted by moving air.

Activity 4.2.2 The Pressure in a Breeze

a) Place an empty can on the table so it's standing upright. Predict what will happen to the can if you blow air straight at the can as shown in the diagram below (viewed from above).

b) Now blow air directly at the can as shown in the diagram above and describe what you observe.

c) Before blowing on the can, we know that the room air is pushing on the can equally in all directions. Based on the above observation, can you draw any conclusions about the forces acting on the can while you are blowing on it? Explain.

d) Using the relationship between force and pressure, can you make any conclusions regarding the pressure of moving air (compared to that of still air)? Explain briefly.

e) Now try the following experiment. Place two empty cans on the table approximately 5 cm. apart from each other. Predict what you think will happen if you blow directly between the cans as shown in the diagram below (viewed from above).

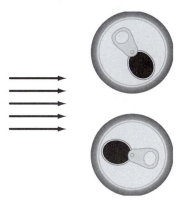

f) Now blow directly between the cans as shown in the diagram above and describe your observations. In what direction do the cans move immediately after you blow? **Note:** This observation works best if you do not blow continually. Instead, give a big, *quick*, puff of air and have your group mates look at the cans from directly above.

g) Based on the above observation, can you draw any conclusions about the forces acting on the cans while you are blowing between them? Explain.

h) Using the relationship between force and pressure, this second experiment should lead to an experimental conclusion about the pressure of moving air (compared to that of still air) that may seem somewhat contradictory to the first. Describe a rule that can distinguish between these two situations. That is, under what circumstances does moving air result in a higher pressure? Under what circumstances does it result in a lower pressure? **Hint:** Try both experiment several times and consider the motion of the air in relation to the objects.

Most students are not surprised by the fact that when you blow directly on an object it moves away from you. However, analyzing this situation in terms of forces and pressures is probably new. More surprising is the strange behavior of the cans when you blow just off to the side. This strange behavior is called the *Bernoulli effect* and has some interesting and counterintuitive consequences. One of the more striking is discussed in the following activity.

Activity 4.2.3 Lifting by Blowing

a)   Take a one-holed rubber stopper and insert a straw (or small, rolled up piece of paper) into one end so that you can blow air through the hole. Try to insert the straw so that it stays in place. Next, punch a thumbtack through the middle of a 3"x 5" card and tape it in place. Now place the card on your hand (thumbtack pointing up) and the rubber stopper/straw on top of the card as shown in Figure D-29. Briefly explain what you expect to happen to the 3"x 5" card when you blow through the straw.

Figure D-29: Insert a straw into the hole in the rubber stopper and place the bottom of the stopper against an index card. Place the stopper over a tack inserted through the index card to keep the stopper centered. What happens when you blow into the straw?

b)   Now try the experiment. Place the card on one hand and try blowing hard through the straw while holding onto the spool with your other hand. Then slowly remove your hand from under the card. If nothing interesting happens, try again and blow harder. Describe what you observe and compare with your prediction.

c)   We know that the force of gravity is acting to pull the 3"x 5" card downwards. What can you conclude about the net air force acting on the card? What does this tell you about the pressure above the card compared to below the card? Explain briefly.

d) Below is a picture of the card with the rubber stopper above it. The distance between the straw and card has been exaggerated. Draw arrows indicating the flow of air in this experiment and explain why this airflow will result in an upward force on the card. Justify your response by relating it to the one of the can experiments.

┌───┐
│   │
│   **Checkpoint Discussion: Before proceeding, discuss** │
│   **your ideas with your instructor!**   │
│   │
└───┘

A Tale of Two Pressures: Moving Air and the Bernoulli Effect

The preceding activities demonstrate that moving air can behave in two very different ways. First, we've seen that when air hits an object directly, the object moves in the direction of the air. This means that the air that hits the object must be creating a higher pressure (compared to the still air on the opposite side of the object), which causes a net force in the same direction as the moving air. Most students don't find this all that surprising. Since air cannot pass right through an object, the air that collides with the object[17] is being compressed somewhat by the air that is coming in from behind, resulting in a higher pressure. Of course, the air will eventually find its way around the object.

Far more surprising (and interesting) is what happens to the air that flows parallel to the surface of an object. We've seen that when air flows along the surface of the object, the object moves into the flow of the air. This means that the moving air creates a lower pressure (compared to the still air on the opposite side of the object), which causes a net force into the moving air. This behavior is much more difficult to understand and the details of why it happens are beyond the scope of this class. Its roots lie in the fact that air moving predominantly in one direction, like wind, does not push the same in all directions like still air does. We have seen that moving air pushes harder in the direction

---

[17] In fact, the air that is actually in contact with the object at the collision point is not moving at all. This point is called a *stagnation* point and the higher pressure is called the *stagnation* pressure.

that it is moving and less push in the directions perpendicular to its motion. As strange as this behavior seems, it is not too difficult to verify experimentally using a pressure sensor and a compressed air source. This is the topic of the next activity.

Activity 4.2.4 Measuring Pressures

a) Begin by setting the computer to take data for about 30 seconds. Then use a pressure sensor to measure the pressure of the (still) air in the room. After about 5 seconds, use a compressed air source to blow air directly into the tube of the pressure sensor. Try changing the air speed by moving the air source further away from the tube. Describe your observations and print out a copy of your graph. What you can conclude about the pressures of moving air and still air in this situation?

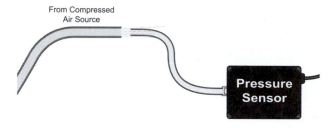

b) Now perform the same experiment, but this time, direct the flow of air across the end of the tube as shown below. This experiment is a little more difficult than the last. You might need to adjust the orientation of the tubes slightly. Describe your observations and print out a copy of your graph. Explain what you can conclude about the pressures of moving air and still air in this situation.

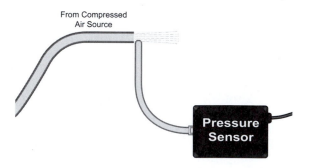

c) Does this data support the assertions we've made regarding the pressure of moving air? Explain briefly.

If you have any problems with the above activity, be sure to talk to your instructor. The results should be very clear.

Airflow and Airplanes

Having learned about the pressure of moving air, we are now in a position to re-visit the question of how an airplane can fly.

Activity 4.2.5 Airplane Wings Revisited

a) Recall that in Activity 4.2.1 you deduced what the pressures around an airplane wing must be in order for the airplane to fly. Shown below is another sketch of an airplane wing. Based on what you've learned about pressure and moving air, determine where the speed of the air should be the largest and where it should be the smallest. Sketch a couple of arrows representing the air speed above and below the wing. **Note:** Longer arrows should represent faster moving air and shorter arrows should represent slower moving air.

b) Explain how this airflow pattern generates the forces necessary to allow an airplane to fly.

c)   Shown below is a sketch of an airplane wing again, but this time,
     the airflow around the wing is indicated by arrows. The longer
     arrows represent faster moving air. Was your answer in part a)
     consistent with what's shown here? Mark the places on this sketch
     where the pressure will be the larges and where it will be the
     smallest.

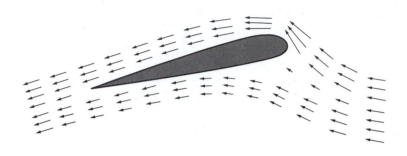

d)   The traction between a car's tires and the road surface is very
     important in racing. Better traction is achieved when a larger force
     pushes the car against ht road. Unfortunately, a heavy car leads to
     poor acceleration and reduced fuel efficiency. Analyze the upside-
     down wing illustrated below and explain how such a wing could be
     used by race car builders to increase traction without significantly
     increasing weight.

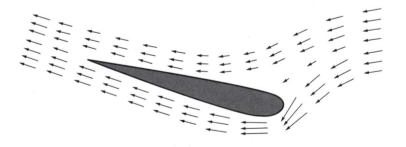

**Checkpoint Discussion: Before proceeding, discuss
your ideas with your instructor!**

As the last activity demonstrates, it is possible to achieve an upward force simply by moving a wing through the air. This is usually referred to as *lift* force. In fact, wing-like structures (called *airfoils*) are used in many different ways to apply forces to objects moving through fluids. One example is the so-called *spoiler* that is used on racing cars to increase the force pushing them against the road without making them any heavier.

Closing Remarks

Throughout this unit, we have developed an understanding of buoyancy, pressure, and airplane flight based almost entirely from experimental observations. We began by trying to understand floater and sinkers which led us to develop a scientific understanding of forces. Then we discovered that it takes less force to lift an object when it is submerged in water than when it is not. We found that this was due to the net upward force of the water on the object —the buoyant force. Then we discovered that the same force could feel quite different depending on the amount of surface area over which the force was acting. This led us to define the quantity pressure as the force exerted divided by the area over which it is exerted ($P = F/A$). We then looked back to see how pressure differences could be used to explain the buoyancy phenomena we investigated early on. Finally, we observed that moving air can lead to pressure difference that can cause forces large enough to lift airplanes!

It may be useful to conclude our study of pressure by comparing two situations we saw earlier in this unit.

Activity 4.2.6 Buoyancy and Bernoulli

a) What do the buoyant forces we studied for sinking and floating objects and the lift forces we studied for airplane wings have in common? Describe how the concept of pressure can be used to explain these two seemingly different behaviors.

We have come a long way from the beginning of this unit. We anticipate that you now have a stronger understanding of buoyancy and pressure than when you began this unit. More importantly, we hope that you have developed a better appreciation for the process of science and the nature of scientific inquiry.

5 *PROJECT IDEAS*

It is now time for you to take on the role of scientific investigator and to design a research project focused on some aspect of this unit that you found particularly interesting. On the pages that follow, you will find a number of project suggestions. Please do not feel limited by these suggestions. You may modify any of these or come up with a completely new one on your own. We have found that many of the best projects are those dreamt up by students. We therefore encourage you to develop your own project on a topic that you find interesting. You should of course consult with your instructor as some projects require too much time or impossibly large resources. Nevertheless, anything involving pressure or buoyancy is fair game. So use your imagination and have fun!

Your instructor may ask you to write a brief proposal that outlines the goals of your project and how you plan to accomplish them. You may find it helpful to refer to the project proposal guidelines included in Appendix B. Try to plan your project in stages, so that if you run into difficulties early on you will at least be able to complete the data collection, analysis, and interpretation. To this end, it is important to note that the project proposals listed here are intended to foster your creativity, not to tell you exactly what to do. In most cases, answering all the questions in one of these proposals would take far more time than you have. So, choose a few questions that interest you or generate some of your own, but try to keep your project focused.

You will probably want to keep a lab notebook to document your project as it unfolds. Also keep in mind that you may be presenting your project to your classmates, so be prepared to discuss your results, how you measured them, and what conclusions you can draw from them. You may find it helpful to look over the oral presentations guidelines and project summary guidelines in Appendix B as you work. These guidelines may give you a better idea of what is expected from a typical student project. Be sure to consult with your instructor about their requirements for your project as they may differ from the guidelines laid out in Appendix B.

Good luck, and have fun!

5.1   SAILING FASTER THAN THE WIND

©PhotoDisc Inc.

Courtesy Bob Dill

Figure D-30: In 1993, Thierry Beilak broke the world sailing speed record on a windsurfer. His record speed was 45.32 knots (83.93 km/hr). While this doesn't seem to be remarkably fast for a land craft, it is incredibly fast for a boat.

Figure D-30: An even more amazing feat was achieved by Bob Schumacher and Bob Dill. In March 1999, they sailed their land yacht (a wind powered land craft) to a speed of 116.7 mph (187.7 km/h) in 25-35 mph (40-60 km/hr) winds.

How can someone sail faster than the wind? While it seems impossible, windsurfers and racing sailboats routinely do it. In fact, people have been sailing against the wind for thousands of years. A close look at modern sail powered crafts gives an important clue as to how one can both sail upwind and faster than the wind. All of the fastest sail powered crafts have sails that are rigid or semi-rigid and wing shaped.

Your task is to explore how a rigid wing shaped sail can improve the performance of a wind powered craft. Begin by constructing your craft and figuring out how to measure the speed of your craft and the speed of the wind. Then investigate the behavior of a simple flat sail. Finally, look at the behavior of a curved/wing shaped sail and compare it to the flat sail. The following suggestions may be helpful.

1.   Using a Pasco cart on a track as your sailing vessel, investigate how a simple land sailing craft might work. Begin by using a flat sail. Using a motion sensor and a pressure sensor/air speed sensor, determine the speed of the cart headed directly down wind with various wind speeds. Explore how changing the angle of the sail and the between the wind and the direction of travel affects the speed. Can you get the cart to sail upwind? Can you get the cart to sail faster than the wind speed?

2.   Experiment with a curved sail? Make the same measurements for the curved sail as you did for the flat sail. How does the performance of the curved sail differ from that of the flat sail? Can you get the cart to sail upwind? Can you get the cart to sail faster than the wind speed?

Other investigations that may be of interest include: (1) The role of pressure in sailing, (2) A comparison of sailing craft and airplanes, (3) The influence of sail size on the maximum speed of the cart, (4) An exploration of other factors that might affect the speed of the cart.

5.2   AIRPLANE WINGS

Figure D-31: The NASA Helios is all wing. This remote controlled aircraft is powered entirely by energy it absorbs from the sun. Unlike the Gossamer Penguin shown at the beginning of this unit, the Helios can fly over great distances and up to altitudes comparable with gas powered aircraft. (Courtesy NASA)

Modern Airplanes carry enormous loads and can fly over great distances. Unfortunately, these aircraft are very inefficient. Recently, NASA and other airplane designers have been working to develop highly efficient aircraft. The fruits of these efforts has been the creation of solar powered aircraft like the Helios shown above and the Gossamer Penguin pictured at the beginning of this unit. These airplanes use the same principles to achieve flight as traditional airplanes. However, by making the aircraft extremely light and by improving the efficiency of the wings, designers have been able to reduce the power needed to keep these planes in flight.

In class we explored the role of pressure differences and Bernoulli's principle in creating lift. In order to develop the low power consumption aircrafts described above, designers had to create wings that would maximize the amount of lift generated while minimizing the power needed. Your task is to explore the variables that affect lift. Make creative use of force and pressure sensors along with a wind source and wings you have found or constructed to research the following questions.

1. What are the variables that affect the lift generated by a wing? For example, these might include wind speed and wing angle. How can you measure lift as a function of these variables?

2. How is the pressure difference above and below the wing related to the amount of lift generated? Devise a way to measure these pressure differences.

3. How do wind speed and the angle of the wing affect the lift generated?

4. In addition to the vertical force of lift, wings also experience a horizontal force called drag. How does drag change as you adjust the variables you identified above? How can you measure the drag forces on your wing?

5.3   HOW DID ARCHIMEDES EXPOSE THE FRAUD?

ARCHIMEDES erster erfinder scharpfffinniger vergleichung/
Wag vnd Gewicht/durch außfluß des Wassers.

©Archivo Iconografico, S.A./Corbis Images

According to the legend of Archimedes, King Hiero commissioned a goldsmith to make him a crown. Heiro provided the goldsmith with a particular weight of pure gold to make the crown. Upon receipt of the crown, Heiro suspected that the goldsmith had replaced some of the gold with the same weight of a less valuable material like silver. In Section 2.3 we read that Archimedes used what he knew about floating and sinking to determine if the crown made for King Hiero was indeed pure gold. According to the legend, he submerged an amount of gold with a weight equal to that of the crown into a container of water that was completely full. He then submerged the crown into the water. Since the crown was diluted with a metal of lower specific weight than gold, it had a larger volume than the pure gold and thus displaced more water. As a result, he was able to detect the fraud.

Some historians doubt the validity of this story. They argue that while it might be possible to use this method for an object made with two materials having dramatically different specific weights, it would not be possible to determine if a third of the gold had been replaced by silver. These historians argue that the uncertainty in measuring the volume of displaced water would be much larger than the difference in displaced volume between the crown and an equal weight of pure gold. Your task is twofold. First, you should test the validity of Archimedes' logic using two materials with dramatically different specific weights. Then you should try to determine whether Archimedes method is a feasible way of detecting the goldsmith's fraud.

1. Use the method of legend (measure the volume of displaced water) to differentiate between two objects of identical weight. One of these objects should be made of pure clay, while the other should have an object of smaller specific weight embedded in the clay.

2. What happens as you reduce the size of the object embedded in the clay? What is the smallest object you can detect with this method?

3. Think of an alternative method that might be more precise. Test this method in the same way you tested the method described in the legend. In this way, find a method that will allow you to detect even smaller objects embedded in the clay.

4. Assuming that the goldsmith replaced 1/3 of the gold in the crown with silver, calculate the difference in the volume of water displaced by the crown vs. the pure gold. To do this, you will need to find the specific weight of gold and silver (look these up or convert from density). You will also need to estimate the size of the crown. Use these calculations to determine if any of your methods could determine the authenticity of Heiro's crown.

5.4   SIPHONS: WATER OVER THE MOUNTAINS?

Courtesy Water Resources Center Archives, University of California, Berkeley

Over the past 20 years, the town of Drysville has been growing rapidly. As a result, the local water resources have become increasingly scarce. In an effort to deal with this water shortage the city council has requested that two civil engineering firms submit plans for obtaining water from other sources. The firms responded, each with a plan for drawing water from a nearby lake.

The first plan submitted by Mountain High Construction calls for running a long pipe from Lake Hyupair (8,000 ft.) over the top of Barrier Ridge (10,000 ft.) down to Drysville (4,000 ft.), a distance of 15 miles. The second plan submitted by Valley View Builders entails running a pipe from Lake Krawstheway (5,000 ft) down into Lolo Valley (1,500 ft.) and up to Drysville (4,000 ft.), a distance of 25 miles.

According to both firms, once the system is set up and the pipes are filled with water, the water will continue to flow by siphon action. They claim that siphoning occurs because there is a pressure difference between the input and the output ends of the pipe and that siphoning will continue as long as the pressure difference is maintained. The city council asks you to assist them in deciding which plan to choose. Your task is to learn how siphons work, decide if the plans presented by the two firms are feasible and choose a favored plan based on your research. Questions you might explore to help you achieve these tasks include:

1. What is a siphon? How does it work? Explore these questions by making a small siphon in the lab. Use what you learned in studying the barometer to develop a theoretical model of how the siphon works

2. What affects the rate of flow through a siphon? How do the cross sectional area and length of the tube (or pipe) used affect the flow rate? How does the difference in height of the input and output ends of the tube affect the flow rate?

3. What are the limitations of Siphons? Are there any conditions in which a siphon will not work? Can a siphon pass over any height of obstacle? Using what you know about barometers, predict what might limit the height over which a siphon can pass. Devise an experiment to test this prediction.

4. Which of the two plans proposed for solving Drysville's water shortage is most realistic? Support your conclusion with evidence from your research on siphons.

5.5   INTO THIN AIR: MEASURING AIR PRESSURE AS A FUNCTION OF ALTITUDE

©Galen Rowell/Corbis Images

Every year, mountaineers struggle to reach the tops of high peaks like Mt. Everest. Although the climbing on Mt. Everest is not technically challenging, very few of those who attempt to climb it actually reach the summit. In many cases, climbers are thwarted by altitude sickness. Typical symptoms of altitude sickness include fatigue, shortness of breath, and headaches. These symptoms are usually mild, however at altitudes as high as Mt. Everest, altitude sickness can become life threatening.

Although you found no measurable change in pressure with altitude in the classroom, altitude sickness occurs because air pressure decreases dramatically at high altitudes. Your task is to measure the change in pressure with altitude. In addition you may also investigate other factors that affect pressure and learn more about the physiological causes of altitude sickness. Use the following questions to guide your investigation.

1. How does air pressure change with altitude? Armed with a pressure sensor, a portable data collector, a topographic map and a GPS unit, devise and carry out an experiment to measure the change in atmospheric pressure with altitude. Use the results of this experiment to estimate the air pressure on the top of Mt. Everest.

2. What causes the air pressure to decrease with altitude? Compare the air in the atmosphere to the water in a container. As you move up from the bottom of the container of water does the pressure increase or decrease? Using what you know about why the pressure changes with depth in a container of water, develop an explanation for why air pressure changes with altitude.

3. Using your pressure vs. altitude data, devise a method to estimate the specific weight of air. Do you think the specific weight of air will change with altitude? Why? Recall that the specific weight of water did not change with depth. Even so, water and air are very different. For instance, water is essentially incompressible while air is easily compressed.

4. If the specific weight of air did change with altitude, how would the pressure vs. altitude graph change. How would this affect your estimate of the pressure on top of Mt. Everest?

5.6   ADJUSTING BUOYANCY: ARE FISH FLOATERS OR SINKERS?

© Steven Hunt/The Image Bank   ©Cousteau Society/The Image Bank

When placed in water, most objects we deal with will either sink to the bottom or float to the surface. Still, there are some objects that don't do either. Fish, divers and submarines all have the ability to float in the middle of the water. Objects that don't float and don't sink are called neutrally buoyant. Any animal or person that wants to be able to move about freely underwater must become neutrally buoyant. As we saw in Activity 2.3.4, when a neutrally buoyant object is completely submerged, it displaces precisely it's weight in water.

At first glance, making an object neutrally buoyant seems simple, adjust its weight and/or volume so that it's weight is equal to the weight of water it displaces. However, things are not quite this simple. Besides maintaining neutral buoyancy, fish also need to be able to move up and down, vertically through the water. In order to do this, they must have some mechanism for adjusting their buoyancy. Your task is build a model fish whose buoyancy you can adjust. Use the following questions to guide you as you construct your model fish.

1. What is required for an object to maintain neutral buoyancy? What has to happen in order for your fish to move up or down in the water? How might a fish or a submarine adjust its buoyancy from being a floater to being neutrally buoyant to being a sinker? Using what you know of Archimedes' principle and the properties of floaters and sinkers, design and build a simple model fish that you can adjust to be a floater, a sinker, or neutrally buoyant.

2. In the ocean, the specific weight of water changes with depth and salinity. How might this affect your fish? If the specific weight of the surrounding water changes, what will be required in order for your model fish to remain neutrally buoyant?

3. How does the mechanism you have used to adjust your model fish's buoyancy compare to the way real fish adjust their buoyancy? How does it compare to the way submarines adjust their buoyancy?

4. Based on what you know about the way a fish adjusts its buoyancy, predict what will happen if a fish is brought from very deep to the surface. Can you simulate this with your model fish?

5.7   THE UNDERWATER WORLD OF DIVERS

Andrew G. Wood/Photo Researchers

We have seen in class that pressure increases as you go deeper and deeper under the surface of water. This raises an important question. Exactly how do our lungs react to this increasing pressure? As a member of Jacques Cousteau's international underwater team, you are commissioned to get information about the way our lungs might behave under the water.

One way of doing this is to build a model lung, take it under the water, and observe how it changes at different depths. Before settling on a model lung, think about the characteristics of your own lungs. Should your model lung be flexible or rigid? Should it be open to the water or sealed? Should you use something like a plastic bag or a test tube?

In order to get some convincing data, your team must devise a way of measuring the changes in size of your model lung as it is taken to different depths beneath the surface of the water. You must also be able to describe how these size changes depended on the depth under the water. The following questions may assist you in further exploring the effects of pressure on divers.

1.   Construct a model lung from a bag or other appropriate container. Devise a method for measuring the volume of your model lungs, the pressure in your model lungs and the depth of your model lungs. Measure the volume and pressure vs. depth in a local swimming pool. How do the pressure and volume change as a function of depth?

2.   Devise a method to make the volume of the model lung remain constant as the lung is lowered deeper into the water. Measure the pressure as a function of depth while lowering your "constant volume" model lung into the pool.

3.   How does the pressure in the water around the diver change as she goes deeper and deeper into the water? If a diver holds her breath while descending the amount of air in her lungs will not change. Will the pressure in her lungs change as she descends? What will happen to the volume of her lungs?

4.   If she takes a breath from her regulator while deep in the water and holds it as she rises to the surface, the amount of air in her lungs will not change as she ascends. How will the pressure in her lungs change? How will the pressure in the surrounding water change? What will happen to the volume of her lungs.

APPENDIX I: USEFUL PHYSICAL QUANTITIES

CONVERSIONS

Length

1 m = 39.37 in = 3.281 ft

1 in. = 2.54 cm

1 mi = 1.609 km

Force

$1 N = 1 kg \, m/s^2$

Pressure

$1 Pa = 1 N/m^2 = 10^{-2}$ millibar $= 1.45 \times 10^{-4} \, lb/in^2$

$1 atm = 1.013 \times 10^5 \, Pa = 1013$ millibar
    $= 14.7 \, lb/in^2$

1 atm = 760 mm Hg = 760 torr

Area

$1 m^2 = 10^4 \, cm^2 = 1550 \, in^2$

$1 km^2 = 10^6 \, m^2$

Volume

$1 ml = 1 cm^3 = 1 cc$

$1 m^3 = 10^6 \, cm^3$

1 gal = 3.786 liters

$1 liter = 10^3 \, ml = 1.0576 \, qt$

PREFIXES

tera	=	$\times 10^{12}$	centi	=	$\times 10^{-2}$
giga	=	$\times 10^{9}$	milli	=	$\times 10^{-3}$
mega	=	$\times 10^{6}$	micro	=	$\times 10^{-6}$
kilo	=	$\times 10^{3}$	nano	=	$\times 10^{-9}$

PHYSICAL CONSTANTS AND PROPERTIES

Acceleration of gravity	g	$9.80 \, m/s^2$
Specific weight of water	ρ_{H_2O}	$1 \, g/cm^3$
Specific heat of water	c_{H_2O}	$1 \, cal/(g \, °C)$
Specific weight of air	ρ_{air}	$1.20 \times 10^{-3} \, g/cm^3$
Specific heat of air	c_{air}	$0.239 \, cal/(g \, °C)$

APPENDIX II: SUGGESTED PROJECT GUIDELINES

PROJECT PROPOSAL GUIDELINES

Your project proposals are meant to make certain that you have done some preliminary planning regarding your project. They actually offer your instructor an opportunity to assess the difficulty of the project you are planning and to help you plan a project that can be completed in the appropriate time frame. Although these proposals are mostly for your benefit, you should adhere to the following guidelines:

Format:

Your proposal should be typed on standard 8 x 11 inch paper. In addition, you should avoid the use of typestyles that make it difficult to read. Typically, a proposal should be one page in length with an equipment list on a separate page. Put your names and project title on all sheets.

Elements to be included in the Summary:

Basically, your proposal should give a reasonable idea of what you plan to accomplish and how. You will not be required to stick completely to the proposal once you begin your project. However, because of time constraints, totally changing the focus of your project is seldom a good idea.

- Brief statement of the purpose of the project.
- A plan for any data measurement techniques you will be using. You should include a list of any equipment you will need to complete your project so that your instructor can gather this equipment before the next class.
- What kinds of graphs you might be making.
- If the continuation of your project depends on preliminary results that you will be making, do your best to explain how you will continue once these results are obtained.
- A project timeline. While your actual experiment may deviate from this timeline, it is a good idea to plan out what you will need to do in the next few class days to complete your project. Planning ahead may allow you to avoid some unpleasant delays.

One final word of advice. Working on your projects is your responsibility. Due to the independent nature of the work, there is a tendency for students to put off the project until the deadline for completion nears. Because there are usually unforeseen problems when attempting any scientific experiment, you are urged to begin your projects early. One of the skills we hope you learn is how to deal effectively with unforeseen (and sometimes difficult) problems.

Furthermore, there will probably *not* be enough class time available for you to complete a substantial project. Thus, you will be expected to spend some time outside of class working on your projects (there is no homework assigned during the projects). In fact, you should use class time to discuss some of the problems you may be having with your instructor. Also keep in mind that you will need to plan a 10 minute group presentation of your project and write a summary as well. Putting things off till the last minute is a sure way to cause you a lot of problems and frustration.

ORAL PRESENTATION GUIDELINES

Your oral presentation is a group effort. As such, it is important that you plan in advance who will discuss each section of the presentation. You are presenting your project to your fellow class mates, so you can assume they have an understanding of the material at the level we covered in class (do not assume too much from your audience or they might not understand your presentation).

Format:

You are allowed 10 minutes to present your project, followed by a 5 minute question and answer session. This is not a lot of time, so you should plan accordingly. Each person is required to give a portion of the presentation, so you may want to rehearse together at least once. At the end of 9 minutes, you will be given a signal that you have only one minute left.

You may draw on the board, use overhead transparencies, make posters, or use whatever visual aids you might need to describe your project. In general, you are *not* allowed to bring out the experimental set-up to demonstrate, we would like you to describe it instead.

Elements to be included in the Presentation:

Although your presentation might not contain all of the elements listed below (or it may contain some that are not mentioned), here are some common features of typical project presentations:

- Brief Statement of the purpose of the project. Remember, no one knows what you have done for your project.
- Description of the investigation, along with background information, if appropriate. The procedure used to obtain data should be stated along with any diagrams or figures, if this is helpful.
- Data should be presented in tables that include units.
- Graphs of data and/or modeling attempts. Ideally, spreadsheets with graphing tools should be used. Be sure to label the axes and use units on your graphs.
- Conclusions based on analysis of the data. This is important!! What does the data tell you? You should interpret, not speculate.
- Discussion of the results. Do your results make sense? What kinds of difficulties did you run into? How might the project be improved?
- Brief conclusion of the project.

Keep in mind that 10 minutes goes by very fast. You may not be able to discuss every aspect of your project in the time allotted. Therefore, you may need to leave out portions that are not critical to understanding the project.

PROJECT SUMMARY GUIDELINES

You should write your project summaries as if a fellow physical science student (one *not* on your project team) is reading it. They should be able to understand exactly what you did and why you did it. In addition, there should be enough detail to allow the reader to re-create the experiment and obtain similar results. Thus, if you devise a unique method for making a measurement, your technique should be described in reasonable detail. The following guidelines should help you to organize your written summaries.

Format:

Your summary should be typed on standard 8 -1/2 x 11 inch paper. In addition, you should avoid the use of typestyles that make it difficult to read. The first page should contain the project title, course name, date, project team members, and author. Next comes the actual summary, which is described in more detail below. Last, you should attach an Individual Performance Assessment form.

Elements to be included in the Summary:

Although your project might not contain all of the elements listed below (or it may contain some that are not mentioned), here are some common features of typical project summaries:

- Brief Statement of the purpose of the project.
- Description of the investigation, along with background information, if appropriate. The procedure used to obtain data should be stated along with any diagrams of figures, if this is helpful.
- Data should be presented in tables that include units.
- Graphs of data and/or modeling attempts. Ideally, spreadsheets with graphing tools should be used. Be sure to label the axes and use units on your graphs.
- Conclusions based on analysis of the data. This is important!! What does the data tell you? You should interpret, not speculate.
- Discussion of the results. Do your results make sense? What kinds of difficulties did you run into? How might the project be improved?
- Brief conclusion of the project.

While there is no length requirement for the project summaries, a typical summary might be two or three pages (double-spaced), not including data and graphs.

INDEX